Depth Psychology and Climate Change

The Green Book

Edited by Dale Mathers

Routledge
Taylor & Francis Group

LONDON AND NEW YORK

First published 2021
by Routledge
2 Park Square, Milton Park, Abingdon, Oxon, OX14 4RN

and by Routledge
52 Vanderbilt Avenue, New York, NY 10017

Routledge is an imprint of the Taylor & Francis Group, an informa business

British Library Cataloguing-in-Publication Data
A catalogue record for this book is available from the British Library

Library of Congress Cataloging-in-Publication Data

LCCN: 2020911717

ISBN: 978-0-367-23717-2 (hbk)
ISBN: 978-0-367-23721-9 (pbk)
ISBN: 978-0-429-28135-8 (ebk)

Typeset in Times New Roman
by Apex CoVantage, LLC

'This is a timely and visionary book, where depth psychology meets deep ecology. The authors explore, explain and expound solutions to the challenge facing our planet. Contributors are analysts who are also artists, poets, philosophers, professors, sailors, scientists, theologians, historians and activists. They illuminate the climate predicament using shamanism, spirituality, synchronicity, science, intuition and imagination.

Everything is connected. The outer climate crisis reflects the inner crisis of the human spirit. Sickness of the earth is reflected in the sickness of the soul. Personal health and planetary health are two dimensions of one single reality. If our civilisation seeks success through consumerism and materialism rather than fulfilment through community coherence and social solidarity then human spirit is bound to be neglected. If mainstream society views the natural world as an economic resource then the oceans become the plastic sink, the forests become commercial commodity and animals are concentrated in factory farms. Climate crisis is a crisis of greed. As Eric Fromm said, we need to move from "having" to "being". What is the end goal of life? Either we never ask this question or we never have time to find an answer. Faced with climate catastrophe we may be compelled to slowdown and ask this.

This book is a most helpful tool to encourage us to ask such pertinent questions and seek our own answers. As we read we may listen to the small, still, inner voice, which may be a message from our soul and which may also be the message from the universe.'

– *Satish Kumar*

'*Depth Psychology and Climate Change* presents us with a choir of diverse voices, thoughtfully assembled to offer heartfelt and insightful perspectives on our psychological responses to climate change. Highlighting the many ways in which depth psychology can inform – and potentially transform – an uncertain future, the contributors above all illuminate the depth and complexity of our relationship with this beautiful, animate earth.'

– *Sharon Blackie MA, PhD*

Depth Psychology and Climate Change

Depth Psychology and Climate Change offers a sensitive and insightful look at how ideas from depth psychology can move us beyond psychological overwhelm when facing the ecological disaster of climate change and its denial. Integrating ideas from disciplines including anthropology, politics, spirituality, mythology and philosophy, contributors consider how climate change affects psychological well-being and how we can place hope and radical uncertainty alongside rage and despair.

The book explores symbols of transformation, myths and futures; and is structured to encourage regular reflection. Each contributor brings their own perspective – green politics, change and loss, climate change denial, consumerism and our connection to nature – suggesting responses to mental suffering arising from an unstable and uncertain international outlook. They examine how subsequent changes in consciousness can develop.

This book will be essential reading for analytical psychologists, Jungian analysts and psychotherapists, as well as academics and students of Jungian and post-Jungian studies. It will also be of great interest to academics and students of the politics and policy of climate change, anthropology, myth and symbolism and ecopsychology, and to anyone seeking a new perspective on the climate emergency.

Dr Dale Mathers is a Member of the Association of Jungian Analysts, London, UK.

To climate activists everywhere

Contents

Figures

Acknowledgements and permissions

To our editor, Susannah Frearson, for her warm support and enthusiasm.

To Professor Brian Cox and Orbital for permission to use the lyrics for 'There Will Come a Time' (from 'Monsters exist' London: ACP records, 2018).

To Alexis Wright for her permission to reprint material from Carpentaria, 2006, Artarmon NSW, Australia: Giramondo.

To the Guild of Pastoral Psychology, to reprint an updated version of Bernard Sartorius' paper 'The Golem: An Image for Our Time', given to their Oxford Conference, August 2013, and reprinted with the kind permission of the author and the trustees of the Guild (Guild Paper No. 315).

To David M. Black, for permission to reprint 'The Moons of Jupiter' from *The Arrow Maker* (2017) London: Arc Publishing.

To Carola Mathers, for the cover illustration, 'Big Sun' (2020). www.alternativeartsales.com

About the authors

David M. Black MA is a poet, author and psychoanalyst. Has published seven collections of poetry, most recently *Claiming Kindred* (2011) and *The Arrow Maker* Arc Publishing, (2017). Fellow of the British Psychoanalytic Society, editor of *Psychoanalysis and Religion in the 21st Century* (Routledge 2006), and author of *Why Things Matter: The Place of Values in Science, Psychoanalysis and Religion* (Routledge 2011). Website: dmblack.net.

Joe Cambray MD PhD is President/CEO of Pacifica Graduate Institute, California; past President of the International Association for Analytical Psychology; former U.S. editor of *The Journal of Analytical Psychology*; former faculty member of Harvard Medical School, Department of Psychiatry; and past President of the C.G. Jung Institute, Boston. A Jungian analyst, living in Santa Barbara, USA. Publications include: *Synchronicity: Nature and Psyche in an Interconnected Universe*, *Research in Analytical Psychology – Volume 1* (edited with Lesley Carter).

Jules Cashford MA is a member of the Association of Jungian Analysts, London. Studied Philosophy at St. Andrews, winning a Carnegie Fellowship to read Literature at Cambridge. Translated *the Homeric Hymns*, Penguin Classics, (2003); publications include: *The Moon: Symbol of Transformation*, Cassell Illustrated (2003); and *Gaia: From Story of Origin to Universe Story*, Gaia Foundation Press (2010). Co-author, with Anne Baring, of *The Myth of the Goddess*, Penguin (1993). Co-editor, with Tom Singer and Craig San Roque, of *Ancient Greece, Modern Psyche*, Routledge (2019), and contributor to *Ancient Greece, Modern Psyche*, Routledge (2015, 2017).

Grant Clifford MA has been an artist and poet for as long as he can remember. He grew himself up entwining art and nature in Scotland: the relationship persists. Prize-winning post-graduate student at Jordanstone College of Art, Dundee. Studied at the Royal College of Art London. Winner of national and international prizes for painting. Lecturer in Fine Art, Dundee University where student needs initiated his training as a Transpersonal Psychotherapist. In private practice beside the Tay Estuary. Researches the relationships between Creativity, Psychotherapy and Health.

John Colverson MA in Integrative Psychotherapy, and MA in Jungian and Post-Jungian Studies (Essex University). A member of the Association of Jungian Analysts, London, and a keen sailor. Publications include: 'Anorexia and Alchemy' in *Alchemy and Psychotherapy*, Routledge (2014).

Jeffrey Kiehl PhD is a Jungian Analyst with the C.G. Jung Institute of Colorado and the Inter-Regional Society of Jungian Analysts. Adjunct Professor at the University of California, Santa Cruz and adjunct faculty member of Pacifica Graduate Institute, where he teaches Ecopsychology. Studied climate change for forty years at the National Center for Atmospheric Research, USA. Publications include: *Facing Climate Change: An Integrated Path to the Future.* Lives in Santa Cruz, CA.

Ann Kutek MA is a member of the British Jungian Analysts Association. Early in life Ann, an urban child, was dropped into a summer harvest in a faraway country, where the technology was the scythe and the horse. Later, she heard of Jung. After learning to read at Oxford University, she opted to become an analyst, occasional writer and accidental historian, adding to her original identity as a painter. She continues to search and attempt to discern what we do to each other and our habitat. Publications include: 'If Not Now, When? Let's Work on Climate Change' in Kiehl, E., (Ed.) *Copenhagen 2013 – 100 Years On: Origins, Innovations and Controversies*, Switzerland: Daemon Verlag (2015); (with Spike Bucklow) 'Lapis Lazuli and the Blue Soul' published as: 'Le Lapis Lazuli comme symbole de transformation dans la pensée de Jung' in *La Revue de Psychologie Analytique*, no. 4. Autumn 2015.

Dale Mathers MB BS is a member of the Association of Jungian Analysts, London. A former psychiatrist, he teaches analytical psychology in the UK and Europe. Directed the Student Counselling Service at the London School of Economics and was a Mental Health Foundation Research Fellow at St. George's Hospital, London. Publications include: *An Introduction to Meaning and Purpose in Analytical Psychology; Vision and Supervision; Self and No Self; Alchemy and Psychotherapy*, from Routledge.

Rosie Mathers MSc is a post-graduate in Medical Anthropology from University College, London. For the last five years, she has worked for grassroots initiatives engendering social change and improving health and well-being for local communities. She is an active member of Extinction Rebellion.

Chris Robertson MPhil has practiced psychotherapy since 1978; co-founded Re-Vision (1988), a UKCP recognised psychotherapy training; and co-created 'Borderlands and the Wisdom of Uncertainty' (BBC documentary, 1989). Chair of the Climate Psychology Alliance, developing psychotherapy's relevance to the climate crisis. Publications include: co-author of *Emotions and Needs*, Oxford University Press (2002); 'Dangerous Margins' in the anthology *Vital Signs*, Karnac (2012); *Ecopsychology's Wilding*, PPI

(2013); *Well-being of Misfortune: Accepting Ecological Disaster*, CPA (2015); co-editor of *Transformation in Troubled Times*, Transpersonal Press (2018).

Susan Rowland PhD teaches at Pacifica Graduate Institute, California and writes on Jung, literature, ecocriticism and the arts. As part of a project on Jungian arts-based research, she also writes mystery novels. email: srowland@pacifica.edu

Mary Jayne Rust MA is a member of the Society of Analytical Psychology, London. An eco-psychologist and psychotherapist, inspired by art therapy, feminist psychotherapy and Jungian analysis. Journeys to Ladakh (on the Tibetan plateau) in the early 1990s alerted her to the seriousness of our ecological crisis; its cultural, economic and spiritual roots. She lives and works beside ancient woodland in North London. Publications include: *Vital Signs: Psychological Responses to Ecological Crisis*, Karnak, London (2011); 'Nature: Truth and Reconciliation' in *Analysis and Activism: Social and Political Contributions of Jungian Psychology*, (Eds) Kiejl, E; Saban, M; Samuels, A; Routledge, (2016). Website: www.mjrust.net.

Andrew Samuels is a member of the Society of Analytical Psychology, London. Formerly Professor of Analytical Psychology at the University of Essex. A political commentator and theorist, he works as a consultant with political leaders, parties and activist groups in several countries, including USA. He roots his work in citizens' lived experiences, and what can be learnt from therapy carried out with political awareness. While he does not disguise his background in progressive and left-wing politics and his commitment to diversity and equality, he celebrates many different takes on social and political issues. Publications include: *Jung and the Post-Jungians* (1985); *A Critical Dictionary of Jungian Analysis* (1986); *The Father* (1986); *Psychopathology* (1989); *The Plural Psyche* (1989); *The Political Psyche* (1993); *Politics on the Couch* (2001). His latest books are *A New Therapy for Politics?* (2015) and *Analysis and Activism: Social and Political Contributions of Jungian Analysis* (edited with Emilija Kiehl and Mark Saban, 2016). website: www.andrewsamuels.com

Craig San Roque MD is a member of the Australia and New Zealand Association for Analytical Psychology, and is a community psychologist, psychotherapist and writer practising in Central Australia and within indigenous affairs. Publications include works in association with Thomas Singer's project on the Cultural Complex. Co-author with Josh Santospirito of the award-winning graphic novel *A Long Weekend in Alice Springs*. His most recent work is in Routledge's *Ancient Greece/ Modern Psyche* (vol 3, 2019), where part 2 of the Persephone trilogy is published.

Bernard Sartorius PhD is a member of the International School of Analytical Psychology, Zurich. He studied theology (University of Geneva), spent some years as a minister (parish and youth work). Trained as an analyst at the

C.G. Jung Institute, Zurich. In private practice since 1974, in Geneva, Lucerne and Zurich. Teaches and supervises in Switzerland, Tunisia, Russia, Georgia, Ukraine, UK and USA. Extensively studied in Egypt, Morocco and (pre-war) Syria, on Islamic mysticism. Publications include many articles about archetypal symbols in contemporary civilisation in *Analytische Psychologie* (Berlin), *La Vouivre* (Geneva) and Spring Publications (Jungian Odyssey Series). Author of *The Orthodox Church* (translated into many languages).

Andy White B Soc Sci is a psychotherapist in private practice in North Devon, an artist and a writer. Raised in Zimbabwe, educated at Rhodes University, Grahamstown, South Africa, publications include: *Going Mad to Stay Sane, a Psychology of Self-destructiveness*, Duckworth (1990); and *Abundant Delicious*, a reinterpretation of the story of Oedipus (2018) His blog, 'Narcissism and the Lost Goddess: The Shadow of Western Culture Explored Through Myth and Fairy-tale,' has over thirty thousand subscribers. http://andywhiteblog.com

Ruth Williams MA is a member of the Association of Jungian Analysts, London. An integrative psychotherapist and supervisor in private practice. Publications include: *Jung: The Basics*, Routledge (2019); Introduction to Special Edition on 40th Anniversary of the Foundation of the Association of Jungian Analysts. JAP Vol. 63 Issue 3 (June 2018); 'Atonement' in *Alchemy and Psychotherapy: Post-Jungian Perspectives*, Ed. Mathers, Routledge (2014); 'Analytical Psychology' in *The Sage Handbook of Counselling and Psychotherapy*, Ed. Feltham and Horton, Sage (2012). Website: www.RuthWilliams.org.uk

'There Will Come a Time'

There are few certainties in science, but one fact of which we can all be certain is that one day we will die.

Our atoms won't disappear, they will return to the Earth. Some will become parts of the living future. But they will carry no imprint, no memories, no knowledge of the pattern once known as you. In five billion years our Sun will cease to shine. Our planet will die in the searing heat, engulfed by the dying star. The atoms once known as you will be ejected out into space. In billions of years they may become parts of new solar systems with their own stories to tell.

The great cycle of stellar death and rebirth offers a sort of limited immortality. Whether that's comforting is up to you. But ultimately nothing will survive. It will all be gone.

In the far future there will come a time when time has no meaning as the universe expands and fades. Our descendants isolated on an island drifting in ocean of dark will watch as the galaxies evaporate away.

How does that make you feel?

Yet something remains in the darkness. An idea. Science is the ultimate exercise of reason. And our reason confirms deep down what we've always known. Whether human or star, life is precious and fleeting.

We are collections of atoms that can think who discovered this deep truth. We must understand the universe will spend an eternity in darkness after a brief period of light.

Meaning is not eternal. And yet meaning exists today because the universe means something to us.

We must understand that life is precious and fleeting. In doing so we will come to recognise the true value of ourselves, our fellow humans and our civilisation.

The choice before us is not between immortality and eternal darkness. The laws of nature have made that choice. But we do get to choose how long we want to survive. How long do you want the human race to survive?

There will come a time when we're forced to choose: do we destroy our planet or protect it? Do we live together or fight amongst ourselves? Do we expand, explore, do we carry our shared hopes and dreams outwards to Mars and the moons of Jupiter and Saturn and onwards to the limitless stars or do we avert our gaze from the universe beyond and allow all memory of our world will be lost too soon?

Do we close our minds and seek refuge in the ignorant dark of the cave, or do we embrace curiosity and love of knowledge of our fellow humans, of our rare world and of the infinite and wonderful things yet to be known? That time is now.

Professor Brian Cox
(from the album 'Monsters Exist', Orbital,
London: ACP music 2018)

Foreword

Words are tricky creatures, honed and buffeted by culture, and so cannot possibly host the bodily mysteries of an individuating psyche. Jungian analysts can find words tricky. Two ironies pervade: one is Jungian analysts use words a lot; while the bond between patient and analyst is not wholly forged by rational sentences, they do nevertheless constitute a real part of analytic space. To dismiss words and writing as possessing no psychic mystery falls into the trap Jungians are supposed to avoid – splitting off psyche from the body.

Such a potential split takes us to the ironic assumption words propel a person into the least interesting part of the psyche. This is theoretically claimed by the Freudian tradition. In Jacques Lacan's revision of Freud's Oedipus Complex, the infant's dawning understanding of language (growing into appreciation of words) splits the psyche into an irreparable loss of (m)other, body and wholeness. Jung never agreed with Freud's all-encompassing interpretation of Oedipus. The same goes for Lacan and language. For Jung, the unconscious is generative: contributing to potential wholeness and strongly allied with the natural world, possessing a wild fertility deserving respect.

Simply put, the Jungian psyche is skilled with words in a way the Freudian is not. Words can be the engine of consciousness and repress other meanings; these Jung called signs. Words can be ripe with potent energies pointing to the unknown within or without; these Jung called symbols. What lurks in the Jungian symbol is the widest expanse of Jungian ideas, including synchronicity as the meaningful coincidence of inner and outer. Symbols cross the presumed boundary between inner and outer. Archetypes pervade the universe and pour their patterning powers into humans and living beings in many culturally influenced forms; some forms are the words Jung named symbols. In fact, he showed symbols conjoin psyche, body and cosmos. To be recognised as such, they require an accepting attitude, a psychic openness to the stars.

Whether a thing is a symbol or not depends chiefly on the attitude of the observing consciousness; for instance, on whether it regards a given fact not merely as such but also as an expression for something unknown (Jung, CW 6, para. 818).

One of my students is writing her doctoral dissertation on how close Jungian symbols and alchemy are to magic and spells practiced today. It will not surprise readers of this book that what she describes as contemporary spell-craft is reminiscent of analysis of any persuasion. In another rhetorical direction, Jung's work anticipates the trans-disciplinarity of quantum scientist Basarab Nicolescu, who evokes a future academy which recognises multiple realities without proclaiming any as foundational (Nicolescu, 2014). Trans-disciplinarity, like Jung, heralds a reconciliation with several esoteric traditions and marginalised spiritual practices.

In my book *The Ecocritical Psyche* (2012), I suggest the Jungian symbol is a portal between human and non-human nature. Exploring examples from literature, I suggest poetry and novels drawing on the collective unconscious are not 'about' nature but rather are conversations with it. Hence, I suggest that *Depth Psychology and Climate Change: The Green Book* is an architecture of words which extend from page to psyche to cosmos. Words have the potential to communicate beyond the human.

Ecology is a process of learning on a planetary scale what Jung proposed for mental health. As part of a fragile ecosphere, an aggressive part of humankind has tried, foolishly, to control non-human nature, instead of learning to live with it sustainably. Many of Jung's later years were devoted to trying to unpick the dualistic and exploitative aspects of Christian myth. Today, one might reflect upon the deeper pattern of his psychology for an ecocritical approach to myth. While Jung produced a concept of the self which he acknowledged was connected to monotheism, his simultaneous connection with many archetypes in the psyche recalls the multiplicity innate to animism. One version of Earth Mother worship is an animistic society whose members can communicate with rocks, rivers, trees and animals in their spiritual or psychological form.

For Jung, myth was powerful because it mediated between overwhelming powers of the other and the frailty of human consciousness. In our greatest mistakes, myth is implicated, but so too is the possibility of a better relationship with the other, as our nature is indigenous to the planet. Important in the coming trans-disciplinarity is complexity theory; the idea that change occurs because of interactions in incredibly complex systems in nature, and the human psyche-body. While Jung lived too early in the twentieth century for complexity science as such, Jungians such as Helene Shulman and Joseph Cambray point out Jung's collective unconscious resembles a collective adaptive system of archetypes.

Human embodiment co-evolves consciousness with historical complex adaptive systems (Shulman, 1997; Cambray, 2012). In effect, the Jungian idea of individuation, in which archetypal energy is creatively transformed by culturally produced archetypal images, strongly resembles accounts of co-evolving complex adaptive systems. Jung's idea of synchronicity as meaningful coincidence adds an additional principle to mechanical causality. It maps effectively onto interweaving complexity, producing spontaneous and creative new connections.

Jung's archetypal unconscious operates as a complex adaptive system to support multiple levels of psychic being and knowing. It provokes individuation, a dissolving and remaking of consciousness every day, as happens in dreams. Individuation is an evolution of consciousness as an *act of integration* with the body, as well as within the culture. Individuation and evolution of consciousness operate, amongst other ways, through reading: literary images spark into archetypal images. To regard reading as complex adaptive system at work with the ecosystem recalls the vision of Dionysus as the god who is torn apart and remade. Complexity evolution also is a form of Dionysian dis-membering and re-membering of being, in the open system of being human.

The typical rendering of the Anthropocene as the new geologic age assumes a rational, wholly knowable human history, innately separated from nature. This colonial cultural construct has intervened into the non-human and changed it forever. Such is not the vision of complexity science, which is of a Dionysian co-evolving dismembering and re-membering through inter-penetrating complex adaptive systems. Dionysus does not do dualistic boundaries. Human culture and non-human nature are inextricable. Dionysian complex adaptive systems mean humans and nature are continually and creatively co-creating, and do so with every breath and heartbeat, in every poem we read or write.

Complexity science, while Dionysian, is not a promise that the climate emergency will be accommodated positively. Perhaps the myth is a communication from the Old One, the Earth Mother, that those who do not worship Dionysus will die by his dismembering Maenads. They tore apart Pentheus, who refused the god. Dionysus has to be worshipped collectively. Understanding of complex adaptive systems moulding planet and consciousness is not enough without collective mobilisation. Dionysus in positive form is the renewed consciousness of bodily instinctual life, when we remember parts as parts of a never fully rational, knowable whole. Dionysus in dark mode is chaos, which is also the abode of complexity. Here we can see *The Green Book* as an invocation towards a positive Dionysian embrace of our sorely tried planet.

Susan Rowland, October 2019, Ojai, California.

References

Cambray, J. (2012). *Synchronicity: Nature and psyche in an interconnected universe*, Carolyn and Ernest Fay Series in Analytical Psychology. College Station, TX: Texas A & M Press.

Jung, C.G. (1953–77). *Except where indicated, references are by volume and paragraph number to the collected works of C. G. Jung*. 20 vol, ed. by H. Read, M. Fordham, and G. Adler, trans. by R.F.C. Hull. London: Routledge and Princeton: Princeton University Press.

Nicolescu, B. (2014). *From modernity to cosmodernity: Science, culture and spirituality*. Albany, NY: State University of New York Press.

Rowland, S. (2012). *The ecocritical psyche*. London: Routledge.

Shulman, H. (1997). *Living at the edge of chaos: Complex systems in culture and psyche*. Zurich: Daimon Verlag.

Introduction

The man bent over his guitar,
A shearsman of sorts. The day was green.
They said, 'You have a blue guitar,
You do not play things as they are.'
The Man replied, 'things as they are
Are changed upon the blue guitar.'
                Wallace Stevens (2015, p. 175)

In this poem, American poet Wallace Stevens describes two responses to a tune –
two realities: green and blue. Depth psychology is not a science, it's an art; like
music – perhaps the blues? Analysts listen to and interpret personal, intimate
songs about the hardest parts of being human – disappointment, separation and
loss. This book could be a blues – 'I woke up this morning, and the world had
changed forever . . .' – the Coronavirus blues. Our human world will never be the
same. We wrote these songs 'before' – you will read them 'after' – in an unknowable future.

As editor, I've been privileged to work with some fine artists: therapists, analysts and anthropologists from Australia, England, Switzerland and the United
States. Each brings their own voice, religious and spiritual beliefs, politics, clinical and life experiences. Some songs you may disagree with, others you may
already know. If you are reading this, you have felt grief from the pandemic and
eco-grief at the catastrophic changes to our climate. It breaks our hearts too, that's
why we wrote this.

'Heart' is what this book is about. We agreed not to write an 'academic text' –
we could have, but it is not appropriate technology. Trying to solve a problem with
an irrational cause by using reason is like trying to play blues guitar by strumming
a bicycle. Trying to argue from 'scientific evidence' does not work with people
who neither know, care, trust nor have any experience of science (Solc and Didier,
2018, *passim*). We talk to a different dimension of experience: the archetypal and
the collective unconscious. They lie behind, below and above reason. They are
where deep change originates.

Depth psychology, drawing on insights from Jung, Freud and their students, inspires us to a profound shift in our collective approach to climate change. Any massive change is a collective problem requiring collective responses and solutions. We offer bridges between insights in analytical psychology and 'green' issues; practical ways to awareness and action, to move us beyond psychological overwhelm. Reviewing eco-psychology, green politics, analytic theory and myths about change and loss draws on our clinical practice. Here are maps (of the mind) and a compass (the heart), placing hope and radical uncertainty alongside – not in place of – rage and despair, which are appropriate when we wake up each morning with more of our world stolen from us while we dreamed. Our songs show how to use your dreams to steal it back.

Climate change is stealing our future. The science is simple: I learnt it fifty years ago, aged 13. When you burn carbon, you get carbon dioxide. This traps sunlight and causes global warming – the 'greenhouse effect.' Warming melts ice, raises sea levels, causes unpredictable changes in global weather, disrupts the ecosystem. When the poorest and hungriest turn to the depleted forests for food, they bring back interesting new animal diseases – like Ebola and Coronavirus (Vidal, 2020). We've lost the protection of a natural boundary between human and animal disease. To a scientist, this is undeniable: to a politician, anything is deniable. As analysts, we understand denial as one of the primitive defences of the self, along with splitting and projective identification.

Deniers use all three primitive defences: indeed, we all use them as selves need protecting to prevent emotional flooding. Simply put, denial is 'it is not happening' (what virus?); splitting is 'if it is happening, then it is *their fault*,' whoever *they* are (as in conspiracy theories . . . the virus came from a bio-warfare lab, is spread by 5G networks, and so on). Psychiatrists call this paranoia. The word comes from the Greek παράνοια (paranoia), 'madness,' formed from παρά (para), 'beside, by' and νόος (noos), 'mind.' 'Para' can also mean 'beyond' – we are 'beside ourselves' when we are 'beyond the mind.'

Projective identification is a little harder to describe, so here is an example. My son, aged 3, was painting in the dining room. My wife and I were in the kitchen. He shouted out, upset. We ran to see what had happened. He pointed at an ugly black blob on his painting. It was spoilt. He said, 'look what you made me do, you naughty Mummy!' Both of us felt a range of feelings – being analysts, this happened fast – shock, rage, anger. We imagined saying. 'but we were next door' . . . then started to smile. Our son had emptied his feelings – shock, rage, anger – whole, into us. He felt acute grief at something spoilt.

'Eco-grief' is a similar feeling, as we see the tragic effects of turning our home into a greenhouse. We empty our fear into each other, we lose scale, we freeze over. Or we go 'look what you've done, you naughty . . . advanced global capitalists, materialist consumers,' forgetting our part. We cook food, heat homes, drive cars, invest in pension funds which invest in . . . advanced global capitalism. When we remember this, eco-grief becomes eco-guilt. Our unconscious sharing of feeling and emotion is an outpouring of the primitive defences of our collective self.

We are deeply uncertain about the future. Yet, in a strange way, it already exists – in our unconscious. It speaks to us in dreams and in imagination, in creativity. Any personal history (song) can be interpreted (performed) any number of ways. Analysis aims to create personal futures in which people sing their self's true song, rather than versions limited by parental, cultural, social or political expectations. In analysis, any 'fact' can be interpreted . . . spun, if you prefer . . . whether the interpretation is faithful to an objective reality depends on the integrity, or lack of it, of the interpreter; or, indeed, agreement about what is 'objective reality.'

Once, there was a cultural agreement that science represented objective reality; previously, in the West, this role belonged to a version of the Christian religion. When the sixteenth century Italian natural scientist Galileo Galilei showed the Sun was centre of the solar system (not the Earth), the Mighty Church displayed political opportunism, shouting 'fake news! fake news!' as in Brecht's play *Life of Galileo* (1986). Political opportunism is a subtle, or not-so-subtle, way of 'adjusting reality' – also called lying. We live at a time when political lying is done more often, and less well, than usual. One gross display of bungling political ineptitude follows another as self-styled 'leaders' lie about what they're doing about the pandemic or climate change.

Some argue politicians live out the archetype of Trickster (Samuels, 1993, pp. 78–102). Presently, politicians seem trapped in a complex where two archetypes work together – Trickster and Shadow. Lying is a traditional political weapon. Themistocles, saviour of Athens during the fourth century BCE war against the Persians, sent his servant to lie to the Persian Emperor Xerxes; telling him the Athenians wouldn't attack his ships next day. Of course, they did, winning the battle of Salamis – and, ultimately, the war. Eventually, due to his kleptomanic greed, Themistocles fled to exile – in Persia. How this feels morally depends upon which side you're on: Greek or Persian (Herodotus, 2006, pp. 491–2). Our political leaders are not as good at lying as Themistocles was.

The thirteenth-century Italian poet Dante Alighieri described their fate in 'the Inferno.' They end up in the Malbowges (evil pockets), deep in Hell: 'a deep trench full of shit where dishonest politicians are imprisoned head down in their own excrement' (Alighieri, 1949, p. 185). His translator, Dorothy L. Sayers, adds:

> Malbowges is, I think, after a rather special manner, the image of the City in corruption: the progressive disintegration of every social relationship, personal and public. Sexuality, ecclesiastical and civil office, language, ownership counsel, authority, psychic influence and material interdependence – all the media of the communities exchange are perverted and falsified, till nothing remains but the descent into the final abyss where faith and trust are wholly and forever extinguished.

Those who understand and accept climate change science view political inactivity as wicked. The Buddhists would say the cause is the 'usual suspects' – greed, hatred and delusion, known as 'the three fires' or 'three poisons' (*akusala-m©la* in Pali, Buddha's language), symbolised as a pig, a bird and a snake. Each cause desire, the root of suffering. They have always been part of being human, and always will be. They are countered by wisdom, generosity and loving kindness. Easy! Well, no, it isn't. We cannot prescribe the antidote, unless we prescribe Buddha – Buddhists often talk about his teaching as 'medicine.' The three fires result from living in Samsara, the world as it is. This can never be the world as we wish it to be. If we're going to bring about political change, then we're going to have to dive into political shit. This requires courage, vision and new ways of thinking. Where will these be found? In the unconscious, because 'they were there all the time.'

Everyone's unconscious speaks in symbols, the language of poetry and dreams. Learning to form and use symbols gives ways to think new thoughts and apply them, to keep looking for creative solutions instead of being flooded by anger and impotent rage. Everyone dreams. In dreams, we rehearse possible solutions to our life problems in a safe space. Ole Vedfelt, a Danish Jungian analyst, writing about dreams, says:

> Symbols and metaphors are holistic expressions that make connections to other experiential modalities than rational thinking, such as imagery, emotions and bodily sensations, which further open other aspects of memory. Like metaphors in poetry, symbols are sensual expressions that interact with our imagination.

> (2017, p. 64)

I began writing this in September 2019. Millions of young people across the world were striking in protest at worldwide governmental inaction. Our children know their future is being actively stolen. This book is for all future people, to whom we, present people, have deep moral responsibility (Wright, 2018). Facing climate change, 'things as they are' means going beyond political lies to psychological truths. Psychology recognises survival patterns are hard-wired in our brains. These patterns, archetypes, form an eco-system. Try this: imagine each brain cell is the size of a large tree. Your head is a forest of seven billion to ten billion trees, each with connecting roots. There are networks, pathways, glades and clearings. There are rivers (blood vessels). From this network, your unconscious mind arises.

Curiously, the number of brain cells is roughly the same as the number of people on Earth: seven billion. It is suggested that by planting only one and a half trillion trees, the amount of carbon dioxide in the air would fall dramatically. A simple solution everyone can understand. But . . . could it happen? Only when there is a change in collective consciousness. And the collective conscious depends on our collective unconscious. We learnt to tune in to nature over our long evolution, or we would not exist. Our collective unconscious is, above all, adaptable. As a species, we faced climate change every time there was an Ice Age. We can do this again. We are going to have to.

How? Though it may seem impossible, a union between 'deniers' and 'believers' is inevitable. When my grandmother died, I was 7. The idea she had gone away and would never come back was too big as, at that age, I could hardly imagine 'next week.' Denial cannot hold in the face of an overwhelming change, such as global bereavement after a pandemic. It might be kind to see 'denial' as an appropriate, normal response to an unimaginable loss. We need to get on with working through the stages of grief, get over having 'the eco-blues.' If we play our songs on a green guitar, we might only play songs of woe. Our blue guitar is depth psychology. We are not a choir, nor are we singing with one voice. We are not 'offering a solution' – so many are needed. Outwardly, the problem is the Earth's temperature spirals upwards: inwardly, it is greed, hatred and delusion. Any solutions, all the writers agree, will include the spiritual.

One text stood out for all of us, Jung's 'Red Book' (2009). Wrestling with staying sane, Jung realised that his conflict was between 'the spirit of the times' – the materialistic, scientific and successful Professor Jung who knew so much, and 'the spirit of the deep' – the spiritual, artistic and hidden Carl, who knew nothing much and had little wisdom. Often people come to Jungian analysts imagining we have 'spiritual wisdom' . . . I know colleagues who fondly imagine they do. We're writing about reality testing and symbol formation. Exploring 'the spiritual' is not a 'cop-out.' The spiritual is a layer of the unconscious, the symbolic. If we see the Earth as conscious, perhaps She is projectively identifying Her feelings back into us, as payback. If we see the Earth as unconscious, perhaps we're taking back our projections of the Shadow. Depth psychology gives a foundation for radical hope and political action because it gives us a language to talk about the formation and use of symbols, with which we can question basic assumptions and the liars who sustain them.

The writers worked independently, yet together. Naturally, similar themes emerged, and these structured the book. We start with things are they are, then look at images and imagination, symbols of transformation and the mythological exploration of the future. The end of each section is marked by poetry. David M. Black, a psychoanalyst, and Grant Clifford, an artist and transpersonal therapist, provide spaces where you can reflect, change tempo. To hear what their symbols sound like, it's best to read them aloud.

Part one: things as they are

Ann Kutek begins with 'An Open Letter to Greta Thunberg' (Chapter 1), as a representative for young climate activists. Drawing on her own youthful experience as a Polish girl raised in England after the Second World War, she noticed the effects of pollution and exploitation in her native land. She amplifies ideas from London Jungian, Roderick Peters, who noted, 'Since the demise of alchemy, the eagle-like ascent of a scientific world view has de-animated matter so thoroughly that for most people the human body itself has become the last refuge of the divine matter,' and argues for a need to put the spiritual perspective back into thinking about our future.

Mary Jayne Rust, in 'Finding the Eye of the Storm' (Chapter 2) also draws on Peters. A spiritually aware perspective could replace a world view of humans as separate from, and superior to, the natural world – seen as 'resources.' Healing anthropocentrism, our species arrogance, involves taking back projections, re-visioning our animal nature and making visceral connections to the oneness of life, leading to a new sense of human identity and belonging. We need to rediscover ancient gifts, lost in the process of so-called 'civilisation.'

Rosie Mathers, an anthropologist, outlines 'An Anthropology of Climate Change Deniers' (Chapter 3). The greatest obstacle to climate change policy is not science or fact, but people and their fiction. She examines the 'tribes of deniers' and the processes by which they write stories against climate change to delegitimate and derail systemic action. By extending theories which connect societal arrangements to human–ecological relations, she shows how these narratives unfold into multiple claims to truth. Such fictions are a response to scale; the inability of our minds to conceive climate change. We need to ask whether restoring human scale can pave the way to acceptance.

Andrew Samuels encourages us in 'Taking the Green Agenda out of the Margins: Psychological Strategies' (Chapter 4) to question why climate change slips down or slides off Western political agendas and gives 'green politics' marginal status. He links up aspects of 'therapy thinking' to climate change, as we need to praise humanity and human artifice, not bury it in eco-guilt. Guilt never produces change. He asks how we could positively advocate the political desirability of the sacrifices needed to achieve sustainability; contrasting this with an addiction to apocalypse which undermines all attempts at change.

Part two: images and imagination

Chris Robertson re-imagines 'Psychotherapy in our 'Cultural Crisis' (Chapter 5), exploring denial and disavowal as defences. Climate change is a natural force, a challenge to exceptionalism and entitlement. The Shadow of our actions brings guilt, shame and psychological disturbances arising from helplessness. Traditionally individual orientated psychotherapies compound the problem of cultural narcissism. Psychotherapists could look out the window rather than into the mirror, recognise and work with cultural symptoms of eco-grief as they appear in their consulting room.

Bernard Sartorius, a Jungian analyst (and beekeeper) from Zurich presents 'The Golem: An Image for Our Time' (Chapter 6), a closer look at a Renaissance story which anticipates the presently unsolvable crisis of contemporary civilisation whose existence depends on permanent economic and technological growth which is destroying nature. Like Dr Frankenstein in Mary Shelley's story, we have built a monster – and, like the owner of a Golem, we need to remove 'the magic words' from it and stop investing psychic energy in 'material blessings.'

Jules Cashford in 'Imagining Earth' (Chapter 7) explores myth and the collective unconscious. Jung gave us a language to understand and collaborate with a new vision of a living Earth embodied in the symbol of 'Gaia,' the original name of the Ancient Greek Mother Goddess. To speak of the 'Return of Gaia' challenges assumptions about an inherent opposition between 'spirit' and 'nature.' Drawing on philosopher Owen Barfield, she suggests consciousness evolves from participation (identity with nature) through withdrawal of participation (setting ourselves apart from nature) to a new way of participating with nature through imagination and engagement (synthesis).

Part three: symbols of transformation

Jeffrey T. Kiehl, an American Jungian analyst and climate scientist for forty years, used active imagination to talk to 'The Green Man' in 'Engaging the Green Man, Breaking Our Spell of Enchantment' (Chapter 8) – a representative of the natural world. He noticed that the more he intellectualised, the further away the Green Man became. We are enchanted by rationality and materialism. We can't rely on our consciousness to face a problem as immense as climate disruption, so, we need to talk with the unconscious and seek advice from the depthless psyche. Dreams and engagement with the imaginal are powerful ways we can listen to the wisdom it holds.

John Colverson takes this further in 'At War with the Natural World: Nature as Other' (Chapter 9). The ecological crisis we face is like an over-eating disorder: endless consumption trying to fill inner emptiness is mirrored by our consumption of the natural world. Consumerism compensates for and denies this emptiness in our collective psyche which shamanism understands as a collective loss of soul. In their perspective, all the natural world is connected. Psychological defences, built up over millennia, need to be overcome. But, as when working with an individual with an eating disorder, healing comes through emotional connection to loss, and recognising the true nature of the feelings of deprivation.

Ruth Williams celebrates 'Our Connection with Animals and the Universe: Psychology, Symbols, Spirituality and the New Physics' in Chapter 10. Animal symbolism is common in shamanic animals and dreams. She reviews different conceptualisations of energy fields from mythology, modern physics and mysticism which enable us to understand our depth connection to each other, and the living beings with whom we share the Earth. She illustrates her chapter with many examples of therapeutic contact between animals and humans.

Part four: myths and futures

Dale Mathers in 'Time, Intuition and Imagination' (Chapter 11) points out that climate has always changed; the problem is in the rate of change, and how our innate resistance to change evokes primitive defences of the self – encouraged by scientific materialism's denial of the spiritual. Connections between the ideas of Jung and the French philosopher Henri Bergson about time and synchronicity suggest that two of Jung's insights are particularly useful in countering psychological overwhelm: archetypes and the collective unconscious. They give us 'time free' experiences; transcendent and spiritual; offering ways to 'be with what is,' a more solid basis for reality testing and building a future.

Joe Cambray explores the '21st Century Unconscious: Altered States, Oracles and Intelligences' (Chapter 12), a contemporary depth psychological approach to ecology that envisions a non-local, distributed psyche. The human psyche is embedded in nature. Let's revise our concept of 'the unconscious' to an interactive field, neither localised nor time-bound, a complex adaptive system (CAS), capable of spontaneous, self-organising transformations producing new holistic forms. In the past, oracles were used to make a 'best guess' about possible futures. To see our own future, we need to listen to the depth of the unconscious.

Craig San Roque imagines 'Persephone's Suicide' in Chapter 13. His symbolic image begins with the coupling of Hades and Persephone, and her despair at the destruction around us, which he illustrates first using text from the contemporary Australian author Alexis Wright, then with an extract from his Persephone trilogy. This began with imagining the Queen of the Underworld arriving in the Australian Outback, and her reception. In this play, she resigns her job as regenerator of the Earth and explains why. His chapter shows how active imagination leads us to different understanding and adaptations.

Andy White amplifies the fairy tale 'The Singing, Ringing Tree' (Chapter 14). It is about a transition from narcissistic self-interest to mature interdependence with the environment. A Prince seeks to marry a proud Princess who sends him off to find the tree, hoping he'll never return. The villain is the King, as expressed by his shadow, an Evil Dwarf, for whom people are a means to an end rather than the means to belong. Through adversity, the young couple lose their hubris, through sacrifice of possessions and their inflated self-image. They attune to the natural world, experiencing themselves as 'part of,' rather than 'apart from.'

We subtitled this 'The Green Book' to acknowledge our collective inspiration by Jung's 'Red Book.' His method, combining dreams with meditation, active imagination, poetic writing and drawing, gives a practical model for a new engagement with the largest political problem the world has ever faced. Our dreams, our imagination and our collective unconscious know ways forward. It is time to trust them.

Dale Mathers

References

Alighieri, D. (1949). *The divine comedy: Hell*, ed. and trans. by D.L. Sayers. London: Penguin.

Brecht, B. (1986). *Life of Galileo*. London: Methuen Student Editions.

Herodotus. (2006). *The histories*. London: Folio Society.

Jung, C.G. (2009). *The red book*, ed. by S. Shamdasani. London: W.W. Norton & Company.

Samuels, A. (1993). *The political psyche*. London: Routledge.

Solc, V. and Didier, G.J. (2018). *Dark religion: Fundamentalism from the perspective of Jungian psychology*. Asheville, NC: Chiron Publications.

Stevens, W. (2015). *The collected poems of Wallace Stevens*. New York: Vintage Books.

Vedfelt, O. (2017). *A guide to the world of dreams*. London: Routledge.

Vidal, J. (2020). "Tip of the iceburg": Is our destruction of nature responsible for Covid-19?' *The Guardian*, Mar. 18, 2020. Available at: www.theguardian.com/environment/2020/mar/18/tip-of-the-iceberg-is-our-destruction-of-nature-responsible-for-covid-19-aoe [Accessed 27 Mar. 2020].

Wright, C. (2018). Obligations to future people. Thesis for B.Phil, Oxford University.

Part I

Things as they are

Chapter 1

An open letter to Greta Thunberg

Ann Kutek

My fellow Earthlings, do not ask what the Earth can do for you – ask what you can do for the Earth

Dear Greta,

For your single mindedness and determination on behalf of our planet, thank you. There is something gloriously harmonious in your call to action over this emergency which has wakened so many people of your generation and older ones, too. Despite the complacency and outrageous pillage all around, you and your contemporaries can raise a smile, and that is to me a sign of hope.

As I write to you, I am reminded of my much younger self and how gradually, over decades, I came to the awareness you evidently possess already. I remember thinking how would it be when I am much older and there will be all those thousands of youngsters who are fresh to the world and will need to come to grips with the legacy they confront? What would that legacy contain? It was just a flash in my mind, but then over time I became worried. I have undertaken an examination of some major milestones in my life. Perhaps we could each do this type of audit with benefit.

It started like this: as children we used to go and spend our summer holidays with our grandparents on their smallholding in Poland. In the late 1950s, it had become possible to cross the Iron Curtain from Western Europe. We were born and raised in London, but there in the countryside, near Wojnicz, it was like before the Second World War. Although we stayed in a big wooden house, a bit like those you have in Sweden, and they had electricity, there was no piped water indoors. You had to carry it in, ice cold in pails, from the artesian well outside, listening to the clanking chain and heavy metal bucket as it was wound down many metres to hit the water in the echoing pitch-blackness. If you wanted hot water to wash, you had to heat it on the wood-fired range in the kitchen which got going in the morning and was still warm at bedtime. The toilet was down the yard beyond the stable, right next to the dung heap guarded by a giant walnut tree. You had to watch you did not fall into the cow pats or worse still, fall through the great hole in the wide wooden plank which served as the seat in that dark smelly place full of cobwebs and creeping things. Toilet paper was an issue: often the shop in the village ran out (as did the whole country) and then there was just shredded newspaper. If you had trouble walking, as my grandfather later did, he used a bottle or a chamber pot and the help had to carry it down to empty in the privy. Can you imagine how hard ordinary life could be then in Central Europe? It is still so in many parts of the world today. In case you were wondering, there was no fridge, only a north-facing larder, off the kitchen. The only telephone there had been came in with the German Wehrmacht (army) in the Second World War when they threw the family out of the house. They took the cables with them before the Russians arrived to take over the building.

At harvest time, everything was done by hand, from scything, binding the bales with straw and then when they had dried off, piling them onto the horse-drawn cart before doing the back-breaking threshing on the earthen barn floor. There were no tractors or machines, and the bread we ate was spongy and grey inside a thick suede-like crust. With freshly churned butter, it was bliss and you could feel your insides applauding at the first bite. Then there were the fruit trees and soft fruit bushes. You had to wait. Some varieties were early, others late, but worst of all were the pears and walnuts. They only came into season long after we were back at school in England, so we had to make do with the previous year's supply kept in the musty cellar under the house.

The views from the veranda were spectacular. The ground descended 300 metres to the narrow main road and, beyond the fields and farms on the plane, was an enormous dark stretch of forest which filled most of the horizon. However, to the right in the distance were thin smoking chimney stacks, maybe 14 kilometres away in a place called *Mościce*. There they made what we call agro-chemicals, mainly for export to the rest of the Warsaw Pact. So, will you be surprised that one year when we arrived, our 'summer nanny' announced there was no longer any birdsong in the forest? We also heard lots of local people were ill with cancer or had died of lung disease since last year. These were the first alarm bells and earliest inklings I had that the smoke-spewing industry could be connected to collapsing biology. It made me want to weep for the spoiling and the insult. Eventually, my grandparents, other family members and friends all succumbed with the passage of time to frightening diseases, so all that 'fresh air' and wholesome food had not helped – but then, you have to die of something.

During those hot summer months long ago, two figures entered my life with lasting consequences. The first, I believe, will be no stranger to you, Greta. It was *Pippi Longstocking* (Lindgren, 1945): my aunt brought me a translation of Astrid Lindgren's work, and no book had ever made me laugh out loud as this one did. Here was a really impossibly strong little girl living with a horse, on her own, with two friends next door, a boy and a girl, a lot like imaginary friends. She had a sea-faring father – my father was also a sailor – and there was no mention of mother. Some of us do need a break from her. So, I figured there was no reason not to plan my life as independently as Pippi had done, and as you seem to be doing. At that time, I was not detained by what we would now consider politically incorrect aspects of the stories. The message was clear: girls can have their own special ambitions, just as boys are supposed to do. Luckily, I attended a mixed school.

The other figure was far more complex and is taking years to unravel. We were sitting round the table outside one evening with grandfather and visitors discussing things that go bump in the night. We mostly laughed it off, well the adults did, when Uncle Wacek, who worked at that agro-chemical factory, said: 'not so fast; there is a psychiatrist in Switzerland, a serious researcher, who is looking into these phenomena and is prepared to see meaning in the sudden appearance of say, a white horse walking alongside you or the occurrence of unidentified flying objects in the sky.' I did not catch the name of the Swiss doctor, or maybe Uncle Wacek did not mention it. Either way, this riveting contrarian view stuck with me ever since. Only much later was I able to identify the Swiss researcher as Carl Jung. He would have been alive, just, when this conversation took place – but more of him later. As for the white horse, our 'summer nanny,' a local middle-aged woman, told us how one autumn evening she was returning home from church walking the four kilometres on an unmade road in the falling dusk when she felt unease. From nowhere, a full-sized white horse came towards her, bridle-less, and walked beside her between the darkening fields until she reached a copse by her hamlet, when it branched off to the left and disappeared. She reached safety without further incident and felt appeased.

Back in London – this would have been the early 1960s – I heard of the disappearance, many years earlier, of an English officer, the explorer Col. Percy Fawcett (Fawcett and Fawcett, 1953). He went to the Amazon forest to look for tribes of indigenous people untouched by European civilization. He never returned from his last expedition, even though search parties were sent after him and there were alleged sightings. His fate remains a mystery to this day and it inspired me both to become a user of public libraries, now disappearing, and along with many others to think of the Amazon rainforest, especially the Matto Grosso region, as a last major bastion of nature with which Fawcett was so absorbed and which absorbed him. Today, as you know, the populist Brazilian government, under new fascist leadership, is bulldozing the vast area as fast as it can in the name of economic development and Brazil's self-determination. It echoes the stirrings on a much lesser scale of certain European states, which also confuse collaboration in the name of survival and security for being 'told what to do by foreign unelected officials.' Strangely, these objectors to a common cause appear to be predominantly self-important men in leadership positions. Where are the women in all this? Mrs Rouseff, the elected president of Brazil, was recently removed for corruption by people at least as corrupt as she was claimed to be.

A final milestone in my pre-adolescent youth, was the publication in 1962 of a book by an American marine biologist, a woman. She was called Rachel Carson, and her shocking book, *Silent Spring* (1962), was on the front cover of the weekly children's magazine I used to read and warned nearly six decades ago how, unless we took seriously our devastation of nature, there would come a day when birds would run out of habitat and food and would no longer sing to announce the arrival of a new spring. Do you think leaders of industry and politics or humanity in general have paid attention since? (See Atwood, 2012.) Greta, you know the answer yourself. The present response seems to be either to rush away from what is fearsome and uncomfortable and shout loudly against it or, rush into panic mode and shout at those held as responsible for the destruction of our environment. Neither is, in my opinion, a productive reaction to fear. Both rely on a balance of prevailing consensus or fashion and do not address what we each inwardly need to think our particular role and responsibility is in this collective crisis.

I am reminded of what US President John F. Kennedy said in his inaugural address, at around this time: 'And so, my fellow Americans: ask not what your country can do for you – ask what you can do for your country' (Kennedy, 1961). Is this not an injunction each generation needs to be taught and consider in relation to the planet?

I remember reading philosopher Richard Wollheim on Freud's distinction about man being a 'horde' rather than a 'herd' animal, albeit less so in advanced societies (Wollheim, [1971] 1981, p. 268). The significance for me was this played a part in my career decisions. My father was a businessman who developed a travel company with contracts for new train and bus routes before moving on to mass air travel. He was originally motivated to bring together families and friends separated by war or migration. However, it was only a short step to diversifying into the mass travel which we now call tourism. Perhaps to his disappointment, I opted against joining the family business and set out on a different path which crystallized only some years later.

At the age of 25, I rewarded myself for achieving a professional qualification with a trip to Peru whose highlight was to be an overnight stay in Machu Picchu – something I had dreamed of from the age of eight having seen at school a picture of the magical Inca mountain hideout. I wanted to experience it before hotel chains moved in to disgorge thousands of tourists, actually not so different from me. I had already worked in Africa and the Caribbean, as an overseas volunteer[1] and intern, but there in Peru, I became overwhelmed by questions like 'what am I doing here?' and 'what justification have I for being here?' They came from feeling viscerally disconnected from the surroundings, yet stung by the abject poverty in the *favelas* and in rural communities, then being showered with rocks in Cuzco while name-called as *gringo* and probably much worse. I felt ashamed of my idle looking. This exciting and fascinating journey was not without peril from abduction, illness and the hideous, often fatal, mudslides from the sides of vertiginous mountains, where precarious roads had simply slid down the valley. And there was the honed art of theft, with knives deftly used to strip back pockets off their moorings or plunged into bulging bags to cascade the contents.

I was ill prepared for this improvised adventure, with only two or three contacts and little useful language. I was equipped with the unwritten licence of Western privilege and the common arrogance of a baby-boomer. More by luck than good sense, I came away relatively unscathed, no doubt enriched and certainly resolved never again to be a 'tourist' in a struggling country, even though I saw myself as a traveller. The journey achieved its manifest purpose and other purposes which had not even crossed my mind, including: rats creeping over us as we slept in the open as they searched out the corn stuffed into our pockets; menus where the only dishes on offer were variations on guinea pig. There were unexpected joys, such as running into officers of the British army near the jungle and getting precise directions to an isolated elderly English missionary priest living two mountain ranges away, by the Apurimac, a tributary of the Amazon river, who welcomed us with open arms. Our only way to repay his generosity was to leave him a wrist-watch, as his was broken. It found its way back to me a year later, a little dusty, in my office in Brixton, South London, through the byzantine, if Catholic, mission-ary network. As yet, I had not arrived at the notion of my own carbon footprint.

So, Greta, scroll on a few years and in midlife I changed career, or rather saw my earlier occupations in teaching and social work as part of the preparation for submitting myself for the arduous training as an analytical psychologist. Now, the Swiss psychiatrist I had heard about as a child assumed a new significance for my work and concerns as I began to study his original take on things and broad-ranging interests, as well as that of others who have drawn inspiration from him (Figure 1.1).

Now Greta, a word of warning about Professor Jung's writing: his style even in translation is often long winded (there is an update in English on its way as I write),[2] and he uses a lot of Greek and Latin expressions, which are a kind of shorthand for educated people of his time, a bit like predictive texting and emojis are for us today, but he is very interesting even when he gets some things wrong,

Figure 1.1 The Peruvian Pot, an original oil painting by Ann Kutek

according to how we might see them today. I believe he is worth the effort, if only to compare the validity of your own creative imagination and thinking with his.

One of these Jung-inspired people is the medical doctor and analyst, Roderick Peters who I happen to know has read all the vast *Collected Works of Jung* and who in 1987 wrote: 'Since the demise of alchemy, the eagle-like ascent of a scientific world view has de-animated matter so thoroughly that for most people the human body itself has become the last refuge of the divine matter' and that sometimes, the only 'containing vessel for transformations is now the analytic consulting room (Peters, 1987, p. 379).

What this abstract and dense but very important statement says to me is we seem to have abandoned our direct relationship with the Earth and the rhythms of the seasons in the so-called 'developed' parts of the world and inserted in its place a terrifying and thoughtless dependence on an arm's-length systematic predictability, as if it were a scientific and therefore fool-proof procedure which leaves little or no room for anomaly or the uncanny. So many of us only have our own bodies and minds as the last hideout of the creative and magical. Yet it is not a generally shared awareness, except when there is trouble. If I had not been one already, I have become a skeptic on most matters.

It does not mean we should get rid of science or the scientific method, but we should apply it more reflectively and still leave room for things like the content of dreams, apparent coincidences, all manner of accidents and especially what we regard as beauty or the uncanny (Jung, 2009). I believe there is no such thing as an accident. Events we like to call accidents, be they good or bad, are the outcome of a build-up we either did not notice or for which there are seemingly no adequate explanations according to accepted rules. I think we as a species have become so big-headed about our certainty in predicting or controlling things accurately through measurement and calculations, using our hand-held devices, we miss blatant realities which stare us in the face.

Listen to what Peters, himself a trained scientist, tells us in relation to a frequent dream content, the image of *an eagle at the crown of the world-tree, warring with the serpent at its roots*. Peters links this *syzygy* (Greek for a married couple) or *coniunctio* (union) with the *soma* (the body), and observes that there are verifiable impacts on the central and autonomic nervous systems and our mental experience of their functioning. He suggests nervous tissue is an early sublimation in matter's inherent tendency to evolve a mind, and the 'I' (the conscious personality) tends to split its self-experience of *prima materia* (primitive mind), along a line between the 'eagle-mind' (associated with the central nervous system), and the 'serpent-mind' (associated with the autonomic nervous system), beyond our control like our beating heart. Further, he suggests the awareness of spontaneously occurring images of the serpent and the eagle (or their variants) helps orient the conscious personality to the inner and outer experience of spirit and matter. So, 'apprehending one's participative relationship with the autonomous and impersonal mysteries which these archetypal (or common) images represent is scary and awing, but it is also accompanied by a sense of belonging, profundity and meaning' (ibid., p. 380).

There are unnoticed things deeply buried in our instinctual (reptilian) selves, which are unleashed especially in group settings, like conflicts or partying when our slim little ego can be besieged by alcohol, drugs or drowning in self-produced hormones (Kutek, 2000).

From Peters' collection of patient dreams, he reports this recurrent threesome: an eagle, a serpent and between them, the dreamer. Their appearance is characterized by tension and conflict; an opposition between hierarchy and substance, between dominance and submission. These features appear to be common to all people. Jung and some of his followers would call these *archetypal* features.

Peters continues (1987, p. 362): 'In general, the serpent symbols in the Judaeo-Christian myth are linked with the "fallen" state of this world, with the devil, with instinctual desires and with the feminine principle.'

The last bit sounds sexist, but let him go on for now (ibid., p. 363):

> Not far below the surface, . . . there is always to be found, like an underground river which springs up in many places, a stream of living symbols . . . they are universal. It is as if these symbols are the best possible way of expressing something of unchanging importance to people. However they may be interpreted, the interpretations are always less complete than the symbols themselves; the interpretations do not continue to appear in people's dreams for millennia, but the symbols do.

Although interpretations of dreams may offer meaning, they are nevertheless both intrusions and of their time and space, and hence they can only be ephemeral. Peters goes on (ibid., p. 365):

> in the Biblical paradise story and in Norse myth the serpent seems to represent the principle of evil inherent in the world, rather than personal sin. And yet these serpentine forces can be supremely [beneficial] if their overwhelming and destructive power can be withstood.

So here again there is hope.

I personally came across such an arresting example of an imprisoned living symbol – of a 'fallen' conflicted state – in analytic work. A patient brought the following dream and has given me permission to mention it to you:

> Aboard ship, some kind of battleship, the dreamer is in the hold.
> There are four elephants chained up in one of the compartments, but it is so small that the elephants are obliged to remain lying down. The dreamer is awe struck and distressed at the plight of such beautiful and majestic beasts so constricted. The first impulse is to unchain them, but then there is no room

for them to stand up. If they did stand up, what would happen to the ship? The dream fades.

This dream can be interpreted in various ways, either personal to the dreamer or impersonally as a recognizable symbol of what is happening to the elephant species at the present time or what humankind is doing to elephants and nature in general. It can also be an expression of a constricted libido or life force, since animals in dreams are taken to be a symbol of this. None of this disturbs the scientific method, which does not itself have anything to say about the type of linking the mind can do and still draw great meaning from it. In this context, the scientific method has obvious limitations.

Here is another example of a reality science cannot fully account for and does nothing at all for the sadness and alarm some of us feel for the loss. Since the 1970s, there has been a catastrophic decline in London's sparrow population (McCarthy, 2010). Estimates vary, but it is said to be of the order of 92%. This once ubiquitous garrulous bird has all but disappeared. Nobody precisely knows why. One British newspaper, the *Independent*, offered a prize of £5,000 for a peer-reviewed study which could conclusively establish what led to the near disappearance of sparrows from our gardens and streets. The prize is still waiting to be awarded. Suggestions range from rapacious cats and magpies invading cities, to the use of decking and concrete in gardens. But then why have they gone from parks such as Kensington Gardens where the landscape has apparently not changed?

One persuasive reason could be the decline in the invertebrates sparrows need to feed to fledglings during a narrow window of development, or they die. If the invertebrates are scarce, is it the lawn treatments that have caused it or more general pollution? Was there a particular virus or some other vector? How then are other garden bird species surviving? The cheerful sparrow is relatively unimportant in the grand scheme but a recently spotted correlation between the rise in new diesel car registrations and the collapse of the sparrow could be a signifier of trends that have a bearing on the rest of us (McCarthy, 2018), something Jung would call a symbol.

A further example of Jung's encompassing way of thinking is presented by my analyst friend, Leslie de Galbert (2007). In her article on the risk associated with 'eating images,' she says engaging with any sensory experience has consequences for the body and the soul, without the intervention of external mood altering substances. She quotes Jung (see Jung, 1928–1930 [1984], quoted in Bucklow and Kutek, 2015):

> The value of an image for man is essentially its emotional content. We gain access to emotion through sensation; once it dwells in the body . . . it's impossible to look at anything without paying a price for it . . .
>
> Before being able to assimilate an image, before it can be admitted as part of oneself, and thereby, perhaps initiating a transformation, we have to agree to consume it, to let it happen, and run the risk of being punished for it! . . . It

means having to suffer its disruptive and destructive aspects, in some cases it means having to suffer quite destructive images and running the risk of being overwhelmed by them.

Years after he abandoned writing in his special secret diary he called *The Red Book*, Jung published a study of tree symbolism entitled 'the Philosophical Tree' when he was deep into his alchemical research. It chimes with Peters' earlier observation about de-animated matter since the rise of science and contains a warning about throwing out the baby (psyche) with the bathwater (metaphysics). Jung says (Jung, CW 13, para.395):

"Only . . . with the growth of natural science, [has] the projection into matter [been] withdrawn and entirely abolished with the psyche. This development of consciousness has still not reached its end. Nobody, it is true, any longer endows matter with mythological properties. This form of projection has become obsolete . . . Nature has nothing more to fear in the shape of mythological interpretations, but the realm of the spirit certainly has, more particularly that realm which commonly goes by the name of 'metaphysics'.

. . . Just as dreams do not conceal something already known, or express it under a disguise, but try rather to formulate an as yet unconscious fact as clearly as possible, so myths and alchemical symbols are not euhemeristic allegories that hide artificial secrets. On the contrary, they seek to translate natural secrets into the language of consciousness and to declare the truth that is the common property of mankind. By becoming conscious, the individual is threatened more and more with isolation, which is nevertheless the *sine qua non* of conscious differentiation. The greater this threat, the more it is compensated by the production of collective and archetypal symbols which are common to all men".

What I think he is saying is we ditch old mythologies as mumbo jumbo at our peril. All we seem to have done now is to fill the resulting void with a shipload of modern myths like 'fake news' and other dangerous excess. This is because our psyches were formed thousands of years ago to contain some certainties alongside very many uncertainties and anxieties. Anxieties are now being exploited on a grand scale by trickery, something Jung had already identified, by the highly effective and instant intrusion of social media. Aside from the high toll in individual mental ill health as a result, the outstanding victim here is honesty. It is all about manipulation for a predicted result, never minding who gets hurt on the way.

There are things we can do about this. It requires attention and study of people who have gone before. Depth psychologists, such as Jung, have taken an interest in the alchemists' symbolic quest and found an energizing and rich seam of coherent analogies to the world we live in. So, the alchemists' pairings of opposites, such as King and Queen, Sol and Luna, Logos and Eros, have an obvious relationship to the masculine and feminine aspects of everyone's psyche. The alchemical view of *hierosgamos*, (a marriage), can translate as the symbol of the amalgam which makes up, among others, a (Jungian) concept of the *Self*. This is defined as the personality of each of us, with our body, mind, spirit, feelings and political outlook/activity.

The union of opposites, whether conceived of as a psychic or as a chemical conjunction, may result in a release of energy which cannot be contained, whether by the ego or by a chemical container. Then the outcome is not a creative product but a literal explosion, as in dynamite or the fusion reaction in a hydrogen bomb. We can add to this list the many apparently accidental 'ecological disasters' such as Three Mile Island, Bhopal, Chernobyl, the Gulf of Mexico explosion or Fukushima, and so on. (For more about humans exploding, see Redfearn, 1992.) The analytical psychologist, Brian Skea states (2003, p. 335),

> even if overt destruction does not ensue, the product of such a union is not necessarily creative but may be a hermaphroditic monster, as [depicted] in some alchemical drawings . . . or a delusional system constructed out of a contact with the unconscious in the psychic realm.

On an individual level, this might be the compulsion to yield to an addiction, such as getting stoned – which can overwhelm the adequate functioning of the ego and break down the Self. The reconstruction/healing of a shattered self is possible, but the circumstances cannot be guaranteed.

You can see here an analogy between the ordinary Self and the planet as a Giant Self.

Another delusional collective system seldom touched on by depth psychology is the prodigious devastation wrought by conquest and colonialism – greed across the globe, on all continents, involving the parasitism of enslavement and inducing the extinction not only of cultures, but of entire communities of peoples who were in a sustainable *coniunctio* with nature for 30 centuries or more: in Australia, Papua, the Caribbean and the Americas (see Diamond, 1998). With them have gone languages, music, art and medicine, and their fauna, but especially their respect for *dreamtime*. (For more on this topic, see Craig San Roque, Chapter 13 of this volume.) There were and are a few surviving endangered places which have evolved into a sustained relationship between flora and fauna of which their people are a part. One historic example was the Jesuit order's experiment called the 'Paraguayan Reductions' started in the 17th century in what is now the border region of Brazil, Argentina and Paraguay. It is described as a 'socialist theocracy' or 'benign colonialism.' Slavery was forbidden in these distinct rural communities and indigenous peoples' freedoms were respected, except for the required Christian worship. It all stopped with the banning of the Jesuit order in Europe in the late 18th century (see Caraman, 1975). Other still surviving places include the Arctic Circle with its Sami people and an island in the Indian Ocean (a UNESCO

World Heritage site since 2008) which serves as an invaluable bank of endemic flora and a human gene pool, and a unique linguistic treasure trove (Morris, 2013).[3] It is called Socotra, 'the island of poets.'

So, Greta, if we each take it on ourselves to reflect on where we have come from and what we can and wish to contribute to the planet, while we have the chance (Kutek, 2015). There are many ways of doing this, like getting involved in local community issues (Kutek, 2016). It needs some of us to slow down and think. There will be people at hand, like the other contributors to this book, and all the thousands of people young and old you have inspired, who share a similar openness and will to collaborate with people of courage like you in this most important and difficult joint enterprise we have probably ever faced in consciousness as a species.

Good luck to you, Greta, and to your contemporaries,
Ann Kutek

With help from Titus, my canine companion.

Notes

1 I was a teacher in Burundi under the British Voluntary Service Oversees scheme.
2 Professor Sonu Shamdasani and The Philemon Foundation are preparing a new English translation of CG Jung's Collected Works.
3 The British-based charity, Survival International, is dedicated to help sustain the way of life of indigenous minorities throughout the Globe.

References

Atwood, M. (2012). Silent spring, 50 years on. *The Guardian*. Available at: www.theguardian.com/books/2012/dec/07/why-rachel-carson-is-a-saint.

Bucklow, S. and Kutek, A. (2015). Lapis Lazuli and the Blue Soul. In *French translation, published as: Le Lapis Lazuli comme symbole de transformation dans la pensée de Jung' La Revue de Psychologie Analytique*, no. 4. Autumn. Available at: www.revue-pa.com.

Caraman, P.S.J. (1975). *The lost paradise – The Jesuit republic in South America*. London: Sidgwick and Jackson.

Carson, R. (1962). *Silent spring*. London: Penguin Classics.

Diamond, J. (1998). *Guns, germs, and steel: A short history of everybody for the last 13,000 years*. British ed. London: Vintage.

———. (1998). *Germs, guns and steel*. New York: W.W. Norton & Company.

Fawcett, P. and Fawcett, B. (1953). *Exploration Fawcett*. London: Phoenix Press (2001 reprint).

Galbert de, L. (2007). Manger les images du coté de la dévoration. *Cahiers Jungiens de Psychanalyse*, 124, pp. 23–37 (in French).

Jung, C.G. (1928–1930 [1984]). *Dream analysis. Notes of the seminar given in 1928–1930*, Bollingen Series XCIX, ed. by W. McGuire. Princeton: Princeton University Press, pp. 12–13.

———. (1953–77). *Except where indicated, references are by volume and paragraph number to the collected works of C. G. Jung*. 20 vol, ed. by H. Read, M. Fordham, and G. Adler, trans. by R.F.C. Hull. London: Routledge and Princeton: Princeton University Press.

———. (2009). *The red book – Liber Novus*, ed. by S. Shamdasani. London and New York: W.W. Norton & Company.

Kennedy, J.F. (1961). *Inaugural address*. Records of the White House Signal Agency. John F. Kennedy Presidential Library and Museum, Boston, Massachusetts.

Kutek, A. (2000). Warring opposites. In: E. Christopher and H. Solomon, eds., *Jungian thought in the modern world*. London: Free Association Books.

———. (2015). If not now when? Let's work on climate change. In: E. Kiehl, ed., *Copenhagen 2013–100 years on: Origins, innovations and controversies – Proceedings of the 19th congress of the international association for analytical psychology*. Einsiedeln: Daemon Verlag, pp. 709–724.

———. (2016). A Jungian Spoke in the town and country planning wheel: It's the alchemy, stupid! In: E. Kiehl, M. Saban, and A. Samuels, eds., *Analysis and activism: Social and political contributions of Jungian psychology*. London: Routledge.

Lindgren, A. (1945/2013). *English translation: Pippi Longstocking*. London: Puffin Books and Penguin Random House.

McCarthy, M. (2010). Mystery of the vanishing sparrows still baffles scientists 10 years on. *The Independent*, Aug. 19.

———. (2018). Riddle of vanishing sparrow "solved". *Daily Mail*, May 7. Available at: www.dailymail.co.uk/news/article-5697861/The-riddle-vanishing-sparrows-SOLVED. html.

Morris, M.J. (2013). The use of 'veiled language' in Soqo? poetry. *Proceedings of the Seminar for Arabian Studies*, 43, pp. 239–244. JSTOR 43782882.

Peters, R. (1987). The eagle and the serpent. *Journal of Analytical Psychology*, 32(4), pp. 359–386.

Redfearn, J. (1992). *The exploding self – The creative and destructive nucleus of the personality*. Wilmette, IL: Chiron Publications.

Skea, B. (2003). Jung, Spielrein and Nash: Three beautiful minds confronting the impulse to love or destroy in the creative process. In: J. Beebe, ed., *Terror, violence, and the impulse to destroy: Perspectives from analytical psychology*. Einsiedeln: Daimon Verlag.

Wollheim, R. ([1971] 1981). *Sigmund Freud*. Cambridge University Press. Reprinted from Fontana Modern Masters Series.

Chapter 2

Finding the eye of the storm

Mary Jayne Rust

It was in the early 1990s when I started to *feel* the urgency of our environmental crisis. The facts I had known for some years became a lived experience when I visited Ladakh, a traditional society on the Tibetan plateau, in northern India. Like many other many traditional societies, they considered the impact of all their decisions and actions on the next seven generations, taking care to remain in balance with the land which supported them. The influence of Western culture has inevitably arrived (albeit relatively recently), affecting their lifestyles as well as their relationships and inner worlds. In this small society, the links between human actions and environmental change were easy to see. For example, in this region of mountain desert, they practise a system of shared irrigation from glacial meltwater; yet they could see the glaciers shrinking.

On my return home, I could see this was a stark mirror for the path my own culture has been taking through for centuries: a gradual disconnection from the rest of nature, from each other, from self. While practical and political actions remain ever more urgent, I learned from those who had walked this path before me that this crisis is calling us to make a shift in consciousness. I wondered what role psychotherapy might play in helping the collective to rebalance. The following story about a Chinese rainmaker from a hundred years ago speaks to this question. Apparently, it was one of Jung's favourite stories.

The Rainmaker Story

There was a great drought in a village in Northern China. There had not been a drop of rain for months and the situation became catastrophic. After many attempts to frighten away the demons of the drought, the rainmaker was called from another province. He asked for a quiet house and there he stayed for three days. On the fourth day clouds gathered and a deluge of rain arrived. When asked how he made this happen the rainmaker said, "I did not make the rain, I am not responsible. I come from another country where things are in order. Here they are out of order, they are not as they should be by the ordnance of heaven. Therefore, the whole country is not in Tao, and I am also not in the natural order of things because I am in a discorded country. So, I had to wait three days until I was back in Tao, and then naturally the rain came".

(Jung, CW 14, para. 604n)

In this chapter, I will explore some questions and reflections arising from this story, such as: how do we return to right relation to with the earth? What is the deep inner work required of us at this time?

Another way of seeing

This story might seem fanciful or simplistic to the rational mind: how could the Chinese Taoist affect the climate by simply waiting for three days until he was 'back in Tao'? Yet to this wise man – and to many indigenous peoples, as well as modern physicists – inner and outer worlds are intimately connected: one affects the other (see Chapter 11 of this volume). At the same time, those peoples would also tell us it takes a lifetime of deep inner practice for one person to create such an effect on their environment. To return to Tao in the midst of a country's disorder is like finding the eye of the storm: it is neither easy nor simple.

What is the Tao? This is very hard to answer. The sixth-century Chinese philosopher Lao Tsu writes: 'The Tao that can be named is not the eternal Tao' (Lao Tsu, 1973). Nevertheless, we cannot resist trying to describe formlessness. The Tao can be thought of as the natural flow or order of the universe; an essence or pattern within the greater whole that keeps the Universe balanced and ordered. It is a non-dualistic principle.

The rainmaker story tells us the Taoist needs to come to the place which is out of order; then he, himself, is out of order. His first step is to let himself *be affected by* the disturbance. This practice is familiar to many psychotherapists: when I listen to my client, I begin to feel their distress in my psyche/body; there begins a natural cycle of feeling into the distress of the other while returning to my own balance, again and again. Jung describes how sometimes we become *infected* by the psychic illness of the patient (Jung, CW 16, para. 163).

Feeling into the disorder of our times

What happens when we move from knowing about climate crisis to letting ourselves be *affected* by climate crisis? Many different feelings might emerge: anxiety, fear, anger, despair, hopelessness or grief. The scale and complexity of the crisis is overwhelming. The wish to turn a blind eye and return to 'normal' is understandable. Yet the climate crisis will continue to disrupt our lives until we are prepared to stop and *feel* the great urgency of the emergency. Only then can we start to digest the situation, to inquire into the depths of the disorder, to know what the next steps might be. What we find is trauma and dissociation.

Dissociation and trauma

C.G. Jung was the first psychotherapist to write extensively about the importance of our relationship with the earth and the consequences of our modern lifestyles:

> Through scientific understanding, our world has become dehumanised. Man feels himself to be isolated in the cosmos . . . No wonder the Western world

feels uneasy, for it does not know . . . what it has lost through the destruction of its numinosities. Its moral and spiritual tradition has collapsed and has left a worldwide disorientation and dissociation.

(Jung, CW 18, para. 581, 585)

He offers a diagnosis of *disorientation and dissociation* to Western culture. We have lost our place, our connection to land, and we have become separated from the world around us.

The South African animal communicator Anna Breytenbach suggests we are suffering from 'separation sickness' (Breytenbach, 2012). While we cannot be truly separate from the rest of nature (try holding your breath for longer than five minutes!), we have withdrawn from an intimate relationship with the other-than-human world, now seen as a 'landscape' or backdrop for human activity. Further, we have distanced ourselves from our animal nature: our senses apparently lead us astray, our feelings must be kept under control; our intuition is branded as irrational nonsense by the scientific, rational mind.

Seeing ourselves through the eyes of those outside of our culture

Sometimes we can only recognise the deep impact this separation sickness has on our psyches by looking through the eyes of others who have grown up in an entirely different culture which is still embedded in the land. A view of white people as they arrive at Turtle Island (USA) is described here by the relatives of author and activist Jeannette Armstrong of the Okanagan tribe: 'My grandmother said, "The people down there are dangerous, they are all insane". My father agreed, commenting, "It's because they are wild and scatter anywhere"'. Armstrong then translates these statements in a much deeper way:

If I were to interpret/ transliterate the Okanagan meaning of my grandmother's words, it might be this: "The ones below who are not of us [as place], may be a chaotic threat in action; they are all self-absorbed [arguing] inside each of their heads". My father's words might be something like this: "Their actions have a source, they have displacement panic, they have been pulled apart from themselves as family [generational sense] and place [as land/us/ survival]".

(Armstrong, 1995, p. 319)

White settlers are a specific group; yet this powerful description is apt for modern culture and speaks to the diagnosis of dissociation and disorientation offered by Jung. The American ecopsychologist Zhiwa Woodbury suggests climate anxiety is, in fact, climate trauma (Woodbury, 2019). He is not the first to suggest we, as a culture, have generations of undigested trauma creating a 'perfect trauma storm' as the global crisis unfolds. This might also account for our inability to respond adequately: someone who is dissociated cannot feel and therefore cannot respond to an emergency. Furthermore, the disturbance of our times is so extensive, so normalised and frequently individualised, making it hard to see its roots (and see Chapter 5 of this volume).

The clash of two stories

The American theologian Thomas Berry writes:

> It is all a question of story. We are in trouble just now because we do not have a good story. We are in between stories. The old story – the account of how the world came to be and how we fit into it – is no longer functioning properly, and we have not yet learned the New Story.
>
> (1978, p. 77)

The 'old story' he refers to is the story familiar to anyone who has been through Western education. Our history is portrayed as an epic, heroic journey from a primitive dark world of ignorance to a brighter world of ever-increasing knowledge, freedom and well-being. This 'progress' was made possible by the birth of human reason and the modern mind; it's all about onwards and upwards. The American cultural historian Richard Tarnas calls this our 'Myth of Progress' (2007, p. 12). He suggests we are undergoing a collective rite of passage, a death-rebirth initiation – but we have no elders to guide us for they themselves are caught up in the crisis, which is larger than all of us (2001, p. 18).

The earth is seen as a set of resources, or objects, for human use. Wilderness, as well as 'wild' human nature, is seen as dangerous, aggressive and out of control, in need of domestication. Peoples associated with wild nature have also been objectified: enslavement of black Africans, oppression of women and the genocide of indigenous peoples are three examples. We take from the earth with no thought for how we give back: we have lost a sense of reciprocal relationship with nature. We see ourselves as separate from, and superior to, the other-than-human world, as if the earth is there to satisfy the needs of humans.

Using a depth psychology lens, we could say we are in the grip of the young male hero archetype, fighting to free ourselves from the terrible mother earth, forgetting that the earth is also the good mother who provides for all our needs (see Chapter 7 of this volume). This psychological attitude is adolescent: arrogant with little gratitude or thought for the mothers' needs, with little awareness of the need to clear up one's mess, a denial of dependency and vulnerability, and inability to hold together 'good' and 'bad' mother.

A great deal of psychological suffering arises from these cultural attitudes. When the earth is seen as a collection of resources for our use, no wonder narcissism is on the rise; when adulthood is defined by wealth and career success, is it surprising that so many young people suffer from mental health problems or turn to crime? When our wild selves are too domesticated, we lose a vital source of life which can lead to depression and an epidemic of addictions. When dependency and vulnerability are seen as a 'weakness', and projected onto women, no wonder there is disgust and shame around our needs, or that women find it hard to value themselves. When we live in such a divided, unequal society, and those in power deny we are on a reckless path towards ecocide, no wonder people feel rootless, untethered, dissociated, displaced, traumatised, no wonder we see a rise in madness in our society.

The Australian Jungian analyst David Tacey draws on Jung's thinking to describe three stages of apprehending the world. In stage one, the land is sacred, and spirits of the earth are seen as real forces. Tacey, who grew up in Alice Springs amongst Aboriginal peoples, describes how, in their cosmology,

> Landscape is a mythopoetic field . . . at the centre of everything: at once the source of life, the origin of the tribe, the metamorphosed body of blood-line ancestors and the intelligent force which drives the individual and creates.
>
> (Tacey, pp. 145–6) (and see Chapter 13 of this volume).

In stage two, where we are now, these forces are seen as irrational, mere projections of the mind upon inanimate phenomena 'usually to be traced back to hysterical ideas or unruly emotions' (ibid., p. 26). While much has been achieved with the scientific, analytic mind, Tacey sees stage two thinking as responsible for the ecological crisis and calls for a stage three in which the world is once again sacred. He writes:

> Stage two thinking lands us in a spiritual and emotional wasteland, in which reason and science have cleansed the world of all projections, leaving nothing left in the world for us to relate to, or form spiritual bonds with . . . No longer sacred, it becomes real estate or 'natural resource' to be used to satisfy egotistical desires.
>
> (ibid., p. 27)

As we enter stage three thinking, Tacey suggests we are becoming open again to transpersonal forces of the earth and world, but we need new cosmologies appropriate to modern times.

Many writers in the field of ecopsychology suggest if we can 're-connect with nature', or 'fall in love with nature again', our destructiveness towards the earth will cease. Others suggest deep healing of human culture is more complex. The American Jungian analyst Jerome Bernstein describes how indigenous cultures live in reciprocal relationship with the rest of nature, whereas Western culture has moved into an age of having dominion over the earth. This is parallel to Tacey's stages one and two thinking. Bernstein suggests that rather than moving into a new story, or going back to something in the past, we need to rediscover the meaning of reciprocity and how it might be in relationship with dominion, in order to evolve beyond modernity (Bernstein, 2006).

To summarise, those of us who have grew up within the Western mindset struggle with two apparently opposing stories. On the one hand, there is a narrative about achieving mastery and success through domination and control of nature, including our own nature. On the other hand, we see the shoots emerging of a new story, which is a long-known ancient story in our bones, about how to work and live *with* nature. The burning question is how we discover and live this ancient story in modern times.

These two stories reflect the ways we move forward into and embrace life versus being fearful of and needing to defend ourselves against it. For a variety of complex reasons, modern culture is increasingly fearful and defensive as seen in the amassing of weapons capable of destroying the earth many times over. So, what is the psychological work needed to come back into balance?

Shadow work

Facing into the shadow, inviting in the disowned, bringing the unconsciousness into consciousness, is vital to coming back into balance. While this is painful, it offers deep renewal. The American Jungian author Rinda West writes:

> This encounter (with the shadow) is sometimes experienced as a process of breakdown . . . in which one has to give up old ideas of the self, particularly heroic ideas, and accept one's flawed, unappealing, and shameful aspects. In this process a person may enter a liminal space in which old adjustments may no longer work but there is no sign of a new outlook. . . . Confrontation with the shadow . . . involves understanding that qualities one had ascribed to the Other are really one's own. Such recognition feels humiliating . . . but then it's possible for that person to withdraw projections and reclaim other parts of the personality that had been abandoned in the course of developing a persona. A process of reintegration can then begin.
>
> (2007, p. 10)

This is a fine description of where we find ourselves. We see this process at work in the struggle with racism, sexism and numerous other oppressions, on individual and collective levels. For example, in my generation and before, it has been common for men project to unwanted feelings and vulnerability onto women, who are then viewed as weak and over emotional, if not hysterical. When a man can own his feeling side, his vulnerabilities, he can then access a more open-hearted and capacity for relationship, both painful and joyful.

The experience we have of working through human oppressions can help us face into anthropocentrism: the human oppression of the rest of nature. The Australian deep ecologist John Seed writes, 'Anthropocentrism . . . means human chauvinism. Like sexism, but substitute "human race" for "man" and "all other species" for "woman"' (1988, p. 35). There are several aspects here: the many ways in which we project onto animals, plants and the earth; our denigration of our animal nature; and how anthropocentrism is linked to other oppression – for example, how people of colour and women are seen as 'closer to the earth'. I will now explore some of these projections and what may happen when they are withdrawn.

Projections onto the other-than-human world

'I love nature' is a commonly heard phrase. We fall in love with place. The British are renowned for worshipping their animal companions and spending many hours tending their gardens. We love charismatic megafauna: creatures such as dolphins, whales and polar bears capture our hearts. Yet despite this proclaimed love, there remains a thick dividing line between us and the rest of nature. This is conveyed from our earliest experience; we are taught that animals do not feel as we do, that their brains are wired differently and that they do not have our superior intelligence. These lies give us licence to continue cruel practices of factory farming, incarcerating animals for medical experiments or mining mountains because the earth is dead; 'it' apparently is simply an object which has no soul. Further, certain creatures who carry particularly negative projections, such as snakes, wasps, spiders (or anyone who poses a threat to humans), worms, slugs and other slimy creatures, or bacteria and viruses. The phrase 'I love nature' clearly excludes much of the natural world.

These negative projections are embedded in our Western language (not in indigenous languages): 'you greedy pig' or 'you lazy/stupid cow'. It doesn't take psychological insight to see these expressions describe humans: we are the most 'greedy' consumers on the planet; we can also be stupid and lazy. We are currently the most dangerous species on the planet.

Our relationship with the other-than-human world is split. Wild nature is feared as dark and dangerous, or idealised as all-beautiful. Yet we have difficulty in holding the tension between conflicting qualities. Perhaps this has not always been the case. In Western astrology, Aries (the Ram), Capricorn (the Goat) and Cancer (the Crab) are symbols for knowing the many-sided parts of the self; here, the animals are neither idealised nor denigrated. The symbol for modern medicine is an image of a snake wrapped around the staff of the Greek physician Asclepius: the snake's venom has the capacity to heal, poison or expand consciousness.

Eagle-mind and serpent-mind

The British Jungian Analyst Roderick Peters explores such tensions in modern-day dreams and ancient stories in his paper, 'The Eagle and the Serpent: the Minding of Matter' (1987) (and see chapter 1 of this volume). He writes:

> A man in his 30's felt a jarring disharmony between his cool, philosophical states of mind and his descents into hot blooded sexuality. In his dreams he was usually on a raft, trying to get across a river or a lake, and all the time

snakes wriggled up through the woven reeds which he was mortally afraid of. All the time he was aware of an eagle perched on the top of the mast, which seemed unconcerned with the goings-on below. . . . A woman who had no confidence at all in her intellect dreamed of an eagle whose feet were stuck to a rock so that it could not fly. Talking of the dream she said that the rocks need the eagle to fly because otherwise the rocks dreams were earthbound.

(pp. 359–60)

Peters kept a record of many similar dreams and describes how the eagle and serpent stand for archetypal polarities in an age-old struggle. He describes 'eagle-mind' as an experience of getting an objective overview, with piercing, focused vision, a place of transcendence, associated with the elements of fire and air. 'Serpent-mind' is an experience of

being very close to, or inside, the dark earth, a power of a deep and inward kind, piercing and paralysing, a subjective participatory experience . . . the realm of blood and viscera. This mind is lodged deep in collectivities. Blood-mind belongs, as it were, to universal blood; if I see someone gashed and bleeding, my blood-mind is affected almost as if it were "my" blood. It is as if there were no boundary between me and that wounded person . . . it is an activity of the ancient mind.

(ibid., pp. 364–5)

He continues:

The experiences of one's bodily self which come when "I" consciousness allows itself to descend into a participatory awareness are our real connection to the past; we can go down and down through the unending evolutionary layers within our bodily nature, and feel a sense of linking up with the dim-mest and deepest roots of life. Through it we can know renewal, as if we have touched vitality itself. The descent feels full of dangers because we know we have gone into the power of the old serpent. . . . The "I" . . . of our conscious experience . . . is all but submerged in feelings of oneness, oceanic feelings, feelings of isolation, abandonment, eternity, infinity, fear, love, hatred, rage; all the passions in fact.

(ibid., p. 373)

Peters describes how historically, both eagle and serpent are to be found in the world tree, the eagle perched on the branches and the snake curled at the roots, in relationship. Later, the eagle was depicted as killing the snake, as the victor, and the snake has mostly disappeared. In many churches today, the eagle is on the lectern carrying the Bible, and the snake is held firmly by its claws, or absent altogether. This reflects the gradual polarisation between the spiritual life and instinctive life.

These descriptions illustrate our fear of regressing down the evolutionary scale into the 'primitive' and losing our special human qualities; being swallowed by something much larger and more powerful, experiences we cannot rationally explain, being taken over by the senses: a few of the many ways our relationship with the earth might propel us to transcend into the sky.

This portal into oneness and boundlessness is what offers the key to deep healing. This has always been known by many indigenous cultures who have offered (and still offer) many and varied forms of ritual as safe containers, as well as guides for this profound journey (see Chapter 9 of this volume).

Coming to know the other-than-human world more deeply

Withdrawing projections goes hand in hand with knowing the other, letting go of pre-conceived ideas, to listening and learning. The American Jungian Analyst Clarissa Pinkola Estes describes how:

> In the old country, where my parents came from, my father said the big trees were "Guardian Trees". All villages had Guardian Trees at the beginning of the road leading to the little houses. The sound of the wind in the leaves of the trees or the needles in the evergreens would change if people were coming from a distance, walking or on wagon or on horseback. The Guardian Trees would tell when someone was far off but advancing toward the village, and from which direction, and on what kind of conveyance, and sometimes whether they were armed or not.
>
> (Pinkola Estes, 2010)

This description reminds us we once had a more intimate relationship with the other-than-human world, brought about through attentive listening, watching and being open to relationship with all living things. Now a growing community of people seek to rediscover these old ways of knowing. This can start by simple practices of awareness. Here is my experience of being in my local forest:

> When I cross from urban tarmac into forest the air is filled with bird conversations and the rustling of trees; the smell of leaves permeates my nostrils and in every step my senses are awakened. What petty concerns might have occupied my mind drift away as I sink more deeply into my body. I am, for the time being, transformed. As the trees and I exchange breath, I begin to see there is no sharp dividing line between my skin-encapsulated "self" and the rest of nature. The small "I" is now in relation to the larger self, things are back into perspective.

'Dod yn ôl at fy nghoed' is a Welsh phrase which means 'to return to a balanced state of mind', but literally means 'to return to my trees'. Here we find traces of our own indigenous wisdom of returning to Tao.

The forest offers many teachings: the art of being deeply rooted in place, the art of shedding skin and drawing in energy for winter, a different perspective on time. Slowing down enables the art of noticing and the wonder of synchronous connections – an acorn drops onto my lap at the very moment when I was feeling the impossibility of renewal. Everyone within the forest community has something to say: squirrels, birds, insects, dogs walking their humans, as well as the infinite blue of the sky. This is not a 'cure all' but an offering to the imagination, a playful way of finding meaning in our world too full of suffering.

Coming back into relation with our human-animal nature

Getting to know the more-than-human world offers rich metaphors and mirrors for the self; we come to know ourselves as human *animals*. As we move down into the body, the constant chatter inside the head falls away and our other-than-rational ways of knowing come to the fore – intuitive knowing, five senses knowing and emotional knowing. These different ways of knowing are described by Jeannette Armstrong of the Native American Okanagan tribe. According to Okanagan teachings, an individual human is made up of four capacities which operate together: the physical self, the emotional self, the thinking, intellectual self and the spiritual self. These capacities are parallel to 'mind' and connect us with the rest of creation. Of the physical self, Armstrong writes:

> We survive within our skin inside the rest of our vast selves . . . Okanagans teach that our flesh, blood and bones, are Earth-body; in all cycles in which the earth moves, so does our body . . . Our word for body literally means "the land-dreaming capacity" . . . The emotional self translates as "heart" . . . emotion or feeling is the capacity whereby community and land intersect in our beings and become part of us . . . We use a term that translates as "directed by the ignited spark" to refer to analytical thought . . . We know that unless we always join this capacity to the heart-self, its power can be a destructive force both with respect to ourselves and to the larger selves that surround us . . . The spirit-self is hardest to describe. We translate (it) . . . as "without substance while continuously moving outward" . . . this self requires great quietness before our other parts can become conscious of it, and that the other capacities fuse together in order to activate something else – which is this capacity . . . this old part of us can "hear/interpret" all knowledge being spoken by all things around us, including our own bodies, in order to bring new knowledge into existence.
>
> (1995, pp. 320–2)

It is startling to see the self being described as so embedded in the earth, inextricably interwoven.

The ecological self

The term 'the ecological self' was coined by the Norwegian eco-philosopher Arne Naess (1988) to describe the capacity to experience the self as expanding beyond our skin. This might happen in several different ways, such as the experience of self which expands outwards, into the land, as Jung describes here:

> At times I feel like I am spread out over the landscape and inside things, and am myself living in every tree, in the splashing of the waves, in the clouds and the animals that come and go, in the procession of the seasons.
>
> (1961, p. 225)

Or it might be the experience of connecting with a self which stretches back and forth in time. The American deep ecologist Joanna Macy and ecopsychologist Molly Young Brown write:

> To make the transition to a life-sustaining society we must retrieve that ancestral capacity . . . to attune to the longer ecological rhythms and nourish a strong, felt connection with past and future generations.
>
> (1998, p. 136)

Jung also describes how his relationship with stone, which was central to his life from such a young age, helps him to connect to a quality of self which is eternal, beyond the narrow confines of the skin-encapsulated ego:

> At such times (of conflict and brooding on God etc) it was strangely reassuring and calming to sit on my stone. Somehow it would free me of all my doubts. Whenever I thought that I was the stone, the conflict ceased. "The stone has no uncertainties, no urge to communicate, and is eternally the same for thousands of years," I would think, "while I am only a passing phenomenon which bursts into all kinds of emotions, like a flame that flares up quickly and then goes out." I was but the sum of my emotions, and the "Other" in me was the timeless imperishable stone.
>
> (1961, p. 59)

Jeannette Armstrong describes an indigenous view of self, making clear that we are interconnected and interdependent with the rest of life, there is no strict dividing line between self and world. Experiences of expanding beyond the skin-encapsulated ego remind us that we are part of the greater whole often bringing about a sense of great peace – an essential antidote for Western alienation and loneliness, so prevalent in our society today.

Connections with other oppressions: racism, sexism

As I mentioned, anthropocentrism is linked with other Western oppressions. Indigenous peoples, Black people and poor people are seen by some as 'animals'

– meaning they are of a 'lower' order, unable to think intelligently, live in primitive, animal-like ways, lazy, dreaming, with no developed sense of self. This stands in contrast to 'civilised' man who is industrious, technologically and emotionally sophisticated, making order out of wilderness, wealth out of wasteland. Women are cast as closer to nature, seen by patriarchy as less able to think, more intuitive, instinctive, and so on – hysterical, over-emotional, vulnerable, incapable of thinking. It is important to recognise these links, to see social and environmental justice as inexplicably linked. Put succinctly:

> When the white man projects his wild animal instinctual self onto black people, he is left like a monochrome print, cut off from his colour and creativity, inhabiting monoculture.
> When a man projects his vulnerability and intuition onto women, he is left in a cut-off, disconnected world of his analytic mind, autistic and unable to relate.
> When a woman projects out her wild animal self, she becomes afraid of the fur on her face, the hair on her body, her flesh, her instincts, her body.
> When we idealise wilderness, we go off in search of our own divinity, beauty and wild mind, flying to unspoilt places in hordes, in search of peace and tranquillity, inevitably spoiling the places we visit.
> When we imprison the self inside the boundary of our skin, we lose our connection to the greater whole, living in separated units, lonely and alienated from life.

Final reflections

I began with 'The Rainmaker Story', which tells of a Taoist who knows the ancient art of bringing a disordered place back into balance through allowing himself to be affected by the disorder and then finding his own way back into Tao. He takes no personal credit for his actions, for he attributes all healing to the Tao; his task is simply to come into alignment and to offer gratitude for any healing that comes.

This chapter has explored how we might feel into our cultural disorder of planetary crisis in order to bring ourselves, and our culture, back into balance. I have been particularly focussing on our relationship with the other-than-human world and the human animal in ourselves as part of the inner work of our times. The consciousness shift that is required of us involves climbing down from our superior position to take our place alongside all others with whom we share our home, acknowledging that we all have a place in the diverse web of life. This might feel humiliating for some, a challenge to the ego. It is a major shift in human identity.

I will conclude with the question arising from many peoples' hearts: where do we find hope? As the crisis quickens, and people start to face into and really *feel* what is happening, it is easy to fall into despair. Many people suspect there is little hope of averting climate change: the best we can do is adapt to the changes and mitigate the worst, preparing for system collapse in the coming decades. This is the conclusion of sustainability consultant Jem Bendell, whose recent paper 'Deep Adaptation' (2018) has gone viral. His readers are at once relieved that he is unafraid of naming what they see as the reality of our situation, while at the same time they accuse him of spreading hopelessness.

Does the rainmaker feel despair and worry about finding hope when asking for the rain to come? I don't know, but my guess is that he does not concern himself with outcomes, or any future preoccupations. He is simply staying with the present, attuning himself to the energetic field around him. Similarly, when a client falls into despair and protests that life is not worth living without hope for the future, the job of the therapist is to hold faith that in exploring the patterns which shape the present a way through will be revealed and the future will look after itself. Tolerating uncertainty is not easy, but this is a practice which offers a way beyond the polarisation of hope and despair. As T.S. Eliot writes, 'I said to my soul, be still and wait without hope, for hope would be hope for the wrong thing' (1959, p. 28).

I find this is helpful for the situation we are facing. The web of life is so very complex, how can I know what is the best solution for life on earth right now? I have faith that the earth is powerful enough to rebalance herself in some way – although this may or may not include the human species. Our task is to attend to the present moment. Beyond the countless practical tasks, there is much to be done: a confusing mix of attending to the devastating losses as well as to the birthing of a new way of being in the world. In this liminal space, between old and new, there is great upheaval and our familiar containers are falling apart. People are anxious and despairing and in a state of great tenderness as the skin of our culture is sloughing off. We need a new form of cultural psychotherapy, which can help us through this mother of all rites of passage. Building community is all.

I also cannot deny there are things I hope for, but it is not hope for some future dream that all will be well on earth. My greatest hope is that we do not walk into the fires and the floods asleep. That as many people as possible can wake up and attend to the crisis as hand. That we can work together and support each other on this tumultuous journey we are on. That we can be tender with one another as the shit hits the fan. That we can keep returning to the present, over and over again, offering gratitude to the earth, over and over again, for all that we are given.

References

Armstrong, J. (1995). The keepers of the earth. In: T. Roszak, M. Gomes, and A. Kanner, eds., *Ecopsychology: Restoring the earth, healing the mind.* San Francisco: Sierra Club, pp. 316–324.

Bendell, J. (2018). *Deep adaptation: A map for navigating climate tragedy. Initiative for leadership and sustainability, (IFLAS) Occasional Paper.* Ambleside, UK: University of Cumbria.

Bernstein, J. (2006). *Living in the borderland: The evolution of consciousness and the challenge of healing trauma.* London: Routledge.

Berry, T. (Winter 1978). The new story: Comments on the origin, identification and transmission of values. *Teilhard Studies*, 1, pp. 77–88.

Breytenbach, A. (2012). *The animal communicator.* (Documentary film), Natural History Unit, Africa. Available at: www.imdb.com/title/tt8842480/.

Eliot, T.S. (1959). *Four quartets: East Coker.* London: Faber & Faber.

Jung, C.G. (1953–77). *Except where indicated, references are by volume and paragraph number to the collected works of C. G. Jung.* 20 vol, ed. by H. Read, M. Fordham, and G. Adler, trans. by R.F.C. Hull. London: Routledge and Princeton: Princeton University Press.

———. (1961). *Memories, dreams, reflections.* New York: Random House.

Lao Tsu. (1973). *Tao Te Ching*, trans. by Gia-Fu Feng and Jane English. London: Wildwood House.

Macy, J. and Young Brown, M. (1998). *Coming back to life: Practices to reconnect our lives, our world.* Gabriola Island: New Society.

Naess, A. (1988). Self-realisation: An ecological approach to being in the world. In: J. Seed, J. Macy, P. Fleming, and A. Naess, eds., *Thinking like a mountain: Towards a council of all beings.* Gabriola Island: New Society.

Peters, R. (1987). The eagle and the serpent. *Journal of Analytical Psychology*, 32, pp. 359–381.

Pinkola Estes, C. (2010). *The dangerous old woman: Myths and stories of the wise woman archetype. Part two. Transcript of interview with Tami Simon, insights at the edge podcast, sounds true.* Available at: www.soundstrue.com/store/clarissa-pinkola-est-s-5257. html.

Seed, J. (1988). Beyond anthropocentrism. In: J. Seed, J. Macy, P. Fleming, and A. Naess, eds., *Thinking like a mountain: Towards a council of all beings.* Gabriola Island: New Society.

Tacey, D. (2009). *Edge of the sacred: Jung, psyche, earth.* Zurich: Daimon Verlag.

Tarnas, R. (2001). *Is the modern psyche undergoing a rite of passage?* Available at: https://cosmosandpsyche.com/essays/.

———. (2007). *Cosmos and psyche.* London: Penguin.

West, R. (2007). *Out of the shadow: Ecopsychology, story and encounters with the land.* Charlottesville, VA: University of Virginia Press.

Woodbury, Z. (2019). Climate trauma: Toward a new taxonomy of trauma. *Ecopsychology*, 11(1), pp. 1–8.

Chapter 3

An anthropology of climate change deniers

Rosie Mathers

Introduction

As I write this chapter, two years after Extinction Rebellion was founded, the international conversation around climate change has become louder. In and amongst the throng can be heard multiple voices: those passionately calling for urgent and immediate change, those condemning activists for their 'disruptive' approach, and those who deny the problem of climate emergency altogether. This chapter looks at the latter, those whose rejection of climate change – strong enough to cause significant roadblocks to progress over the last 30 years – has, in some cases, amplified along with cries for change.

Denial takes many forms. In psychoanalytic terms, denial manifests as a conscious or unconscious repression of events, experiences, or truths too traumatic or frightening for the individual. In order to cope, people often use 'splitting' – psychically cutting off from a traumatic event so it does not interfere with their ability to function. Whilst emotions, sensations, or thoughts around the trauma can be kept at bay for many years, these buried selves have a habit of turning up elsewhere – in negative relationship patterns, addictive behaviours, poor physical health, and so on. The truth, it seems, will always come out – causing damage and destruction on its route to doing so.

Whilst individual denial is indeed a case for analytical psychology, collective denial requires attention to the processes of social value construction. What beliefs and ideologies underpin a culture? Which power structures legitimate or reject certain risks and taboos over others? Socio-cultural anthropologist Mary Douglas and political scientist Aaron Wildavsky (1983) were among the first to note that 'threats' do not arrive predetermined and fully formed into our lives – they are rather chosen according to specific cultural values to reinforce a particular social hegemony. What is seen as 'risky' often does not pose significant long-term or widespread danger. We see this in current discourse around fundamentalist terrorism versus climate change – the former representing a tiny minority of individuals unlikely to have a significant impact on the global population; the latter (highly risky in terms of scale and impact), is continually depicted as less of a threat than the 'war on terror'.

The political advantages of selecting terrorist over environmentalist risk lies in its potential to divide, frighten, and thus control the population, as opposed to an ecological risk which would require mass social unity and cooperation. Thus, if the construction of risk is a social practice which operates as a form of structural control (Douglas, 1966), the collective denial of risk requires a similarly anthropological lens to unpick the origins, motivations, and manifestations of such behaviour (Zerubavel, 2006).

The phenomenon of climate change denial has been approached from many angles by different disciplines – evolutionary biology, sociology, political science, psychology, behavioural sciences, media, linguistics – each providing a differently nuanced perspective on why we deny our current ecological reality. The actors in these stories are myriad: from people in power motivated by greed to citizens unable to compute the scale of the problem in their everyday lives – and manifest differently across divisions in nationality and class. In this chapter, I incorporate perspectives from these varying standpoints, as I aim to synthesise why people continue to deny climate change, and how this undermines current attempts at governmental response and international policy. I begin with a discussion of human-ecological relations informed by contributions from material culture, environmental anthropology, and eco-feminism to show how the natural world has undergone a long process of 'othering' and 'objectification' in the West. I then show how this contemporary imagining of the earth as a servile 'other' impacts current climate change response – as denialists and policy creators remain influenced by the belief the earth is an 'object' with no agency of its own. In addition, I discuss recent headlines from the denial echo chamber which dismiss environmental activism as a marginal, rather than a mainstream concern. After zooming out to this scale, I bring the final section back to everyday lived reality in discussion of emotional denial, which leads us back to the question of intimate psychological processes and how we deal with mass change.

Human imagination in relationship to nature

As other authors in this volume show, humans have a diverse and complicated relationship to nature. All and at once, the environment provides a source of myth and legend (White, Chapter 14; San Roque, Chapter 13), a reflected mirror of our internal world and processes (Kutek, Chapter 1), a site of trauma of the historical past (Robertson, Chapter 5), political hubris (Samuels, Chapter 4), and a site of personal placemaking – safely locating us within a (seemingly) secure material world. Anthropologists have shown how 'space', rather than the scenery against which life plays out, becomes meaningful 'place' – as people invest terrains with social, moral, and spiritual value throughout their lives. Practices which create shared memory (collective storytelling, the building of houses, towns, and sacred monuments) imbue physical landscapes with social meaning and value – creating a sense of ownership, belonging, and 'home' (Basso, 1996; Ingold, 2000; Low and Lawrence-Zuniga, 2003).

Much of this literature was influenced by German philosopher Martin Heidegger's (1971) assertion that 'building' is more than a functional activity which joins brick, mortar, stone, or wood – but also a means of enabling 'dwelling' – the felt sense of safety and situatedness within our physical world. This subtle distinction between function and feeling provides a new binary through which to view human-ecological relations; by separating out the difference between the material world 'out there' and the meanings with which we ascribe it, it is possible to hold two realities in play – the physical landscape of earth, plants, rivers, and mountains, and our collectively imagined one – filled with history, lived experience, significance, and story.

In this section, I trace the hi-*story* behind human-ecological relationships in the West, as a process mediated by political, economic, and structural forces. Ever since we have had 'civilisation' in this part of the world, we have designed ourselves apart from nature. In the writings of Aristotle and Plato, the organisation of human sensory perception illustrates the beginnings of this division. Their preoccupation with sight over baser sense modalities such as smell, taste, touch, and sound shows a philosophical bias towards cognising, processing, and surveying the world around us, rather than acknowledging our sensuous physical intertwining or situatedness in (rather than on) the earth. Privileging the eyes and carving them off from the reality of full body perception, allowed societies to play out the fiction that we are separate (and above) other species, because we equate sight with judgement and truth. We say 'seeing is believing', never 'feeling is believing', or even less likely, 'smelling is believing'! The eyes are the 'window to the soul', enabling us in the role of curtain-twitching neighbour, omniscient evaluator – a watchtower over the world rather than a participant in it.

Similarly, phenomenologist and anthropologist Tim Ingold shows how the human ability to stand on two feet remains a primary distinguisher of ourselves from nature. The early Darwinian delineation between labour of the hands and feet – where the former represents our ability to calculate, use tools, and thus enlarge brain faculties, and the latter is a mere 'pedestal for the rest of the body' (2004, p. 317) – represents what Ingold names a triumph of 'head over heels' (ibid., p. 318). This victory, he argues, is representative of an ideological fracturing between intellect and sensory experience – the hand crafts images from the brain, whilst the foot *feels*, and passes preceptory messages back, from the ground. Ingold advances this material argument by discussing how the slow distancing between foot and earth – through the use of shoes, boots, tarmac, and bitumen – represents an almost military enclosure and discipline of the perceiving self, alienating us from the physicality of the world below. Again, there exists an imagined superiority, or on-top-ness, between the sphere of social and cultural life and the ground upon which it is enacted.

This sensorial distancing and separation from the physical world has become intrinsically linked with civility and morality. In his seminal work, 'The Civilizing Process', German sociologist Norbert Elias (1978) shows how comportment related to violence, bodily functions, table manners, and sexual behaviour – all activities which employ 'base' senses – became sanctioned by post-medieval standards of repugnance and shame. Containment, refinement, and cultivation became a modicum of sociality – behaviours such as spitting, urinating or defecating in public, and overt sexuality were equated with disgust and incivility. Free and unbounded morals, persons, and even lands became synonymous with depravity: as wild, sensual femininity and lush, overgrown wildernesses implied an intrinsic danger; whilst rich, upper class masculinity and enclosed, cultivated lands represented civilisation and culture.

In her book *Wild: An Elemental Journey*, writer Jay Griffiths shows how the Western notion of land-as-sin is etymologically traceable – words such as villain (*villein*), pagan, (*paysan*), and heathen (heath-dweller) – were once apolitical terms for those living outside the city (2006, p. 42). Over time, this physical distance became synonymous with moral detachment; where the town represented a Christian morality and work ethic, and the heath – wild, devilish, and often female forces. The unruly overgrowth of feral and fecund lands was tantamount in sin to female sexuality – both powerful and potentially destructive forces, threatening to overthrow the existing moral order of masculine gentility. As feminist scholar Silvia Federici notes (2004, 2018a, 2018b), mounting attempts to cultivate and control areas of free, unpasteurised land during the 1700s and 1800s were closely linked to the rising witch hunts, trials, and mass executions. These so-called 'witches', Federici writes, were typically lower-class women who benefitted from a system of collective land use, using 'common' land to grow food, materials, and sometimes medicines, obtaining unauthorised status and power on their own. Violence against them symbolised a wider attack on potential political insubordination and socialist values, as well as their threateningly sensual closeness with the environment.

In addition, the witches' intimacy and knowledge of the earth's healing properties symbolised an empirical 'magic' in direct antithesis to the separatist values of capitalism. As ecofeminist Carolyn Merchant writes in *The Death of Nature* (1980), early capitalist thought demanded a social and material detachment from the land in order to control and monetise it. The simultaneous demise of the 'earth-as-mother' and 'earth-as-organism' rhetoric worked together to strip the earth of its sensual consciousness, enabling an increasingly aggressive approach to resource extraction. Merchant writes:

> One does not readily slay a mother, dig into her entrails for gold or mutilate her body, although commercial mining would soon require that. As long as the earth was considered to be alive and sensitive, it could be considered a breach of human ethical behavior to carry out destructive acts against it.
>
> (p. 3)

Thus, a process of othering, or 'objectification', of nature was required to justify new, more vigorous, and expansive levels of felling, mining, and plundering. The earth was no longer a sentient female host, but a servile object – devoid of inherent value, meaningful only in terms of what it could provide and supply. In the same way as woman is considered 'Other' in relation to man's 'Absolute' (de Beauvoir, 1997), the earth became an appendage to man's conceptualisation of himself, determined and differentiated only in relation to human activity. Thus, land and women were simultaneously reified, objects in the eyes of society – which could be used and pillaged without any moral consequences.

By Victorian times, the idea of man as above nature was well established, and widely reflected in art and literature. Thomas Gainsborough's classical portrait paintings used the natural world as an artistic backdrop for his subjects, many of whom owned the land in front of which they were depicted. For example, his *Mr and Mrs Andrews*, in the National Gallery, London provides a perfect example of human (male) superiority. We see man (standing, with a gun), wife (sitting, hands in lap), and dog (gazing lovingly at his master) occupying the foreground, with nature rolling peacefully in the background. The lush grounds and green fields provide the stage upon which greater, human characters are drawn – encapsulating the Victorian ideal of the 'landscape' as a 'backdrop' upon which human activity takes place.

Such domineering ownership of land is reflected in the literature of the time; Thomas Hardy's *Tess of the D'Urbervilles* describes the yielding potential of pastures and farmland, and the use of industrialised technology to tame and bring nature into submission (as Tess is brought to heel by Alec D'Urberville and Angel Claire). In both examples the triumph of 'thought' over 'materiality' is clear, as human consciousness presides and rules over nature.

The globe and miniaturisation

How does earth-objectification manifest today, and what significance does this have in the context of climate change? In this section, I argue the process of objective miniaturisation, or symbolic reduction, which the planet has undergone due to advances in science and technology, reframes our world as a controllable and governable entity – with no internal autonomy of its own. In *The Savage Mind* (1966), anthropologist Claude Levi-Strauss outlines his concept for the social meaning and purpose behind 'miniaturisation' – the rendering of a real-life object into a 'simplified' copy through artistic methods such as paint or model making. Using examples including a ship in a bottle, the delicate lace collar in a Clouet painting, and a child's doll, Levi-Strauss argues that 'miniatures' induce a profound aesthetic emotional response, which comes from their very abstraction (p. 23). In each instance, the giving up of certain dimensions – physical, tactile, or temporal – reduces our typical pattern of understanding, condensing the materiality of an object into a symbolic representation of the whole. This process affords a type of aesthetic pleasure, whereby the viewer feels themselves in a separate reality from, and thus superior too, that which is being signified. Levi-Strauss writes:

Being smaller, the object as a whole seems less formidable. By being quantitatively diminished, it seems to us to be qualitatively simplified. More exactly,

this quantitative transposition extends and diversifies our power over a homo-
logue of the thing, and by means of the latter can be grasped, accessed, and
apprehended at a glance.

(ibid.)

Through distance and reduction, the object becomes smaller and subordinate,
whilst we become larger and superior. An example of this is the practice of map
making: large expanses of land are transformed and condensed into representative
and readable dimensions. In his book *Imagined Communities* (1983), political scientist Benedict Anderson traces the origins of map making from John Harrison's
invention of the chronometer in 1761. This technological development allowed
geographic longitudes, and later latitudes, to be measured – giving rise to a new
system of visualising and surveying the world around us based on quantifiable
axes, grids, and coordinates. Previously unbounded and undocumented terrains –
forests, grasslands, deserts, oceans – could now be reliably portrayed and interpreted, through a system of symbolic miniaturisation.

This process created new parameters for spatial domination and control (Anderson, 1983; Latour, 1987). As Anderson notes, the map is an essential proponent of
colonialism – providing the imagined possibility, strategy, and means to travel and
settle in previously unchartered lands. Large areas were soon conquered and coloured with 'imperial dye' – blank unoccupied terrains became coloured in British
pink-red, French purple-blue, or Dutch yellow-brown (Anderson, 1983, p. 250).
'Dyed this way, each colony appeared like a detachable piece of a jigsaw puzzle'
(ibid.). He continues – geographic reality was reduced and subsumed into nation-
alistic sign. Contemporary maps retain this jigsaw style – composed of brightly
coloured continents and baby blue oceans – a simulacrum now deeply imbedded
into our universal imagination.

Spatial miniaturisation, as cartography, gave rise to a new way of conceptualising and objectifying the natural world, one which facilitated its capture and
control. In the context of climate change, the three-dimensional rendering of the
globe-as-miniature, rather than a map of its surfaces, perpetuates the idea of a
tractable and governable planet. Philosopher Kelly Oliver, in her book *Earth and
World* (2015), illustrates how the first pictorial versions of earth, 'Blue Marble'
and 'Earthwise', taken during the 1970s Apollo missions, helped to create a conceptual distance between people and planet. These now infamous photographs
provided a new perspective of the world – aloof, distant, and microscopic –
something held at arm's length, viewed from a disembodied vantage point.
Humanity had achieved the seemingly impossible, transcending its natural habitat and *capturing* it, through the eye of the lens. Suddenly, our entire planet could
be seen in miniature – a small ball, the shape of a glass marble, which we could
hold in our hand.

Visit any museum shop today and you find mini 3D globe keyrings, sponge
balls, rubbers, children's toys, bedside lamps . . . you name it. The desktop globe,
a staple in any imperialist office (or Hollywood films where the evil villain always

has a model globe on his table), spins around according to our will as we trace our fingers across its curves and travel across its borders, covering vast distances in under a minute. Its smallest rendering is the emoji – a tiny, textable image, half a centimetre across of familiar blue and green. Such scaling down, productive and necessary to modern life in many ways – facilitating international travel and trade, for example – makes our world more manageable, less 'formidable'. However, as I discuss in the next section, it also facilitates an imagined human supremacy which denies our planet its own complexity, agency, or autonomy.

Objectivity in the context of climate change

In his recent work, French philosopher and anthropologist Bruno Latour (2014) traces the scientific history of contemporary 'objectification' of the earth to its roots in Galilean physics. Latour argues that Galileo's radical discovery that the earth moves in orbit around the sun, rather than vice versa, reimagined the globe as an active and essential part of the solar system – an astrological cog in a wider intergalactic machine. From this perspective of interrelatedness, Galileo put to one side notions of an internal climate or biosphere, and focused instead on earth's position, connection, and function within the solar system. This presented a new vision of earth as an objective, scientific fact – one in which, I suggest, human-time became minor and insignificant in relation to the vastness and limitlessness of space-time.

New parameters for calculating time and distance became necessary as our skies transitioned under Galileo from the 'heavens' (property of God) to 'outer-space' (property of science). Modern astronomical units such as parsecs, light years, solar masses, and lunar distance now reflect the enormity of our universe, making our planet, ourselves, and our actions inconsequential by comparison. This reimagined earth was stable in its vastness, reliably protected from human activity by the sheer force of its immensity.

Scientific enterprise gave us new dimensions to measure the earth – placing it outside of the realm of human proximity, sensuality, or influence. Instead, it became a point of scientific inquiry, taking on properties unique to an object – stable, immutable, constant, and without internal significance. Rather than endowed with a natural subjectivity it became inert – denied of its own inner pro-cesses or agency. Human action had no significantly felt consequences because the earth-as-object could not feel, experience, or respond in its own way. Thus, instead of paying attention to the ways in which people and ecology interact, shape, and mutually inform each other, Galilean science let us see the world as a lifeless mass upon which our existence takes place. This putative separation between human action and planetary re-*action* has dramatic significance in the context of climate change. If we conceive that our behaviours have no impact on the natural world, how can we take responsibility for the human-induced process of global warming? How should we respond to the earth's sudden acts of wild, unpredictable agency?

A problem with contemporary ecological responses, argues Latour (2013), is our continued faith in science as the hallmark of objective, rational truth. In the West, science provides our primary ontological framework, explaining everything from the laws of physical attraction to the chemistry of baking a cake. It aims to describe our lives in absolutes – breaking it down into neat sub-atomic particles with largely predictable outcomes and discernable behaviours (the field of quantum physics which deals with randomness and chaos remains mostly outside of the 'accepted' mainstream, and particularly separate from international policy). However, as sociologists and anthropologists have repeatedly shown, science is, in fact, a social enterprise, governed by the same cultural laws and relativity as any other human practice (Kuhn, 1962; Foucault, 1973; Latour and Woolgar, 1979; Shapin and Schaffer, 1985; Latour, 1987; Martin, 1987, 1991, 1994; Shapin, 1994, 1996). Scientific research agendas, methods, findings, dissemination – even language – always develop in dialectical interaction with society, and both consciously and unconsciously imbue socio-cultural ideals and partialities (Latour and Woolgar, 1979; Shapin and Schaffer, 1985).

The need for scientific truth before action is a logical and reasonable reaction to the climate crisis. As Latour (2013) states, contemporary policy operates on a strict basis of facts before values; science *before* politics. Indeed, determining international policy on the basis of scientific authority is a sensible, necessary approach to multiple global issues – providing a safe and effective response to disease outbreak or extreme weather crises. But it is not the only possible approach, nor the only factor to be considered. In the context of climate change this tactic presents two central problems, writes Latour (ibid.):

1   overlooking the political biases constituting science, and
2   consolidating the need to have conclusive scientific evidence before action.

Both contribute to our current environmental stagnation – the need for un-*deniable* truth before action creates paralysis, especially when those who wish to halt ecological preservation replace fact with phony science. These so-called 'denialists' perpetuate the idea of the earth-as-object – reifying the planet as a simple resource for advanced global capitalism.

The denial machine

The 'denial machine' is a term used to describe various high-stakes actors who refute climate science to defend the existing moral order of free market capitalism (Begley, 2007; Monbiot, 2007; Dunlap and McCright, 2011). The machine is made up of the system's beneficiaries: corporate America, fossil fuel corporations, conservative think tanks, politicians, front groups, contrarian scientists, media outlets, and self-designated 'experts' who use uncertainty to muddy the waters around the ecological crisis. Their agenda is continued earthly exploitation through objectification – the commitment to free enterprise and unfettered growth demands mounting resource extraction and fossil fuel use – without any acknowledgement or responsibility of their consequences for our planet (Jackson, 2009).

Specific tactics include: refuting climate science as faulty, inaccurate, or motivated by a corporate green agenda; providing 'alternative science' in articles, newspapers, and academic literature; and using the media to invalidate and marginalise environmentalist efforts. The aim is to manufacture enough uncertainty to cease or delay environmental sanctions; just as tobacco industries used faux science to instill doubt over claims that 'cigarettes are bad' to keep up their profits (Oreskes and Conway, 2010). The goal of the denial machine is to create *debate* – to set up sides for and against – so the focus of attention becomes winning the argument, rather than moving onto what should be done next. 'Divide and rule' is an effective strategy, as mechanisms for healthy ecological reflexivity are weakened and attempts at meaningful political change are paralysed.

Headlines produced by the denial echo chamber vary in tone, content, and ferocity. In America, right-wing broadcasters employ heavily religious language to frame environmentalists as a literal pox on the nation – of whom Greta Thunberg is the self-elected 'saviour'. Media organisations from the *Wall Street Journal* (Baker, 2019; Freeman, 2020) to the website American Thinker (Showalter, 2019a, 2019b) repeatedly used Thunberg's age and innocence to debunk her influence, dismissing her as 'a cute, sincere kidlet who doesn't know very much' (Showalter, 2019b), or a naughty 'climate school skipper' with a deceptively naïve 'I won't eat those peas' face (ibid.). In another article, she is mocked as the 'child savior of the Earth', preaching the 'climate gospel' with a 'Hail Mary pass' to 'the high church of environmentalism' (Baker, 2019).

The irony in these metaphors paints environmentalists as silly and ridiculous, and at worst, sacrilegious. Their apparently religious fervour likens them to a cult: devout, unreasonable, and wholly irrational – refuting the scientific grounds upon which ecologists base their arguments. As the *Wall Street Journal* mocks, a movement which believes in sin, penance, and salvation 'doesn't sound very scientific' (ibid.). And whilst such comparisons undermine environmental action, they subtly reiterate the conservative, capitalist, and human-ecological supremacy of contemporary American Christendom: 'rather than simply bewailing the costs of growth, Old-Testament style, we should be mindful of its redemptive benefits too. Surely we can all say amen to that' (ibid.).

In the UK, where religion is less bound to political nationalism, attacks on environmental activists marginalise in other ways. Following the international rebellion in October 2019, Extinction Rebellion (XR) members were described in the British media as 'drug taking hippies', 'dreadlocked crusties', 'loopy', 'bad parents', 'doom mongers', 'hypocrites' and so on (Bodkin, 2019; Lawson, 2019; Sabey, 2019).

As Canadian sociologist Erving Goffman writes, focusing on difference in terms of appearance, behaviour, or values, is a stigmatising enterprise, resulting in 'social categorisation into an adverse/undesirable group of people who are deemed to be morally bad' (Goffman, 1963, p. 12). Difference equals immorality, providing legitimate grounds to condemn and ignore the actions of a group. Interestingly, miniaturisation is used again here – to reduce the hundreds of thousands of diverse protestors (XR Doctors, XR Teachers, XR Youth, XR Rainbow Rebellion, XR Disabled Rebels) to one single stereotype – a stoned, unwashed, vagrant. Turning an environmental majority (approximately 20,000 people (Gostoli, 2019) marched in a funeral procession which marked the end of the two-week rebellion in October 2019) into an incidental minority, recategorises ecological preservation as an alternative concern.

By symbolically increasing the numbers of those in power – 'the few' – by suggesting 'we' – the common-sense public – are on their side (Bancroft and Heale, 2019; Welsh, 2019), climate activists become 'othered'; a nuisance rather than a necessity. Continual broadcasting of the protests' cost impact and inconvenience to the Metropolitan Police (BBC News, 2019) cast rebels as extreme, disruptive, and outside the law – unlike 'Us' good, conforming citizens.

Even generally agreed-upon neutral or left wing media outlets in Britain such as the BBC, the *Guardian*, and the *Independent* – whilst never overtly denying climate change, systematically undermine climate emergency lobbyists by dismissing their actions as naïve, idealistic, or too ambitious. Recent publications by the BBC focus on the short-term annoyances of reducing emissions, such as 'severe restrictions on flying', 'ruin[ing] thousands of holidays', with dramatic changes to diet, shopping patterns, and ways of life (McGrath, 2019; Regan, 2019). These demands, endorsed by an 'unprepared', 'incohesive' and 'selfish' group of 'radicals' are inconvenient, unrealistic, and over-the-top (McGrath, 2019; Turns, 2019). By levying their attacks on organisations like Extinction Rebellion, rather than climate change, these reports consistently undermine the need for action. The tone is 'business as usual', protecting the rights and lifestyle of the hard-working average Joe, above all else.

This responsible, morally aligned, self-maximising individual should suffer no ills for his zealous commitment to capitalism – his well-earned holidays must not be compromised, his commute must not be interfered with, and he must always reap appropriate rewards in luxury goods, diets, and technology. Of course, the disruption to everyday life caused by protests are highly inconvenient, for which XR repeatedly apologies. This internalised, manic urgency to get to work on time is not born with the individual themselves, but instilled through the capitalist values which shape our labour force.

Denial in practice

Another roadblock to cogent ecological response is continuing to conceptualise human and ecological worlds as two distinct categories. The repeated segregation of these two spheres is made evident in the conceptualisation of 'natural disasters' as an organic and accepted environmental phenomena, rather than something brought about by human action (Oliver-Smith, 2013; McMahan, 2016). Public narratives around 'natural disasters' like flash flooding and hurricanes (both have increased exponentially in the last 30 years due to heavy resource extraction and elevated uses of fossil fuels) concentrate on apolitical meteorological phenomena; the volume of rainfall, speed of wind, scale of area hit, and so on, whilst leaving out the vital contextualisation of each event as part of the human-induced process of climate change. Online reporting on the recent and devastating bushfires in Australia continued to emphasise natural causes – prolonged drought, strong winds, and the normal, if not slightly elevated, conditions of Australian summer – only acknowledging climate change as one of many contributing factors.

Contemporary climate change policy falls short as it fails to embed itself into human worlds (Oliver-Smith and Hoffman, 1999; Oliver-Smith, 2013). As an example, World Health Organisation (WHO) guidelines (2002) for safeguarding public health in the face of 'natural disasters' are tailored to respond to each disaster as an individual episode, requiring specific and singular provisions (Oliver-Smith, 2013). This strategy of disaster management isolates each event as unique moment in time, one which necessitates preparation and response tactics particular to the incident, without the need for collectivising solutions or planning longer-term mitigations. Rather than stringing together environmental catastrophes and coming up with a unified set of policies, this technique of 'lurching from one disaster to another' (Rosenbaum, 2006) reinforces a systemic denial of human-ecological worlds intermingling, as our actions are treated as having no material consequences on the world around us. The result is international policy which is incohesive, ill prepared, and unable to deal with the scale and complexity of the issue.

Another problem with forging collectivist approaches are the competing viewpoints on what to do. In their paper 'Clumsy Solutions for a Complex World: The Case of Climate Change', political scientists Marco Verweij et al., (2006) outline four different social approaches to the environment which result in dramatically different policy responses. To illustrate, the authors make use of group/grid theory – a system of demarcation designed by social scientists to measure different types of cross-cultural sociality (Douglas, 2007). As shown in Figure 3.1, the core differences in social grouping are represented in a two by two table with two axes. The 'group' axis measures the importance of social bonds, how much an individual is motivated by collective belonging. The 'grid' axis measures the importance of ranking and stratification within a society. The axes cross in the middle, drawing four quadrants, each representing a high/low assimilation along each of the values.

In the top right-hand corner, high importance on group belonging and social stratification gives rise to a hierarchical way of life – individuals are motivated by social bonds and organised according to strict social rank. Below, a strong affiliation to the group, but little import on individual ranking creates an egalitarian society where all individuals have the same value. To the left, a low-group score and high stratification results in individualism where 'every man is for himself', and above, a low score for both group affiliation and social status gives rise to fatalism, where people reject any notion of existential belonging or control.

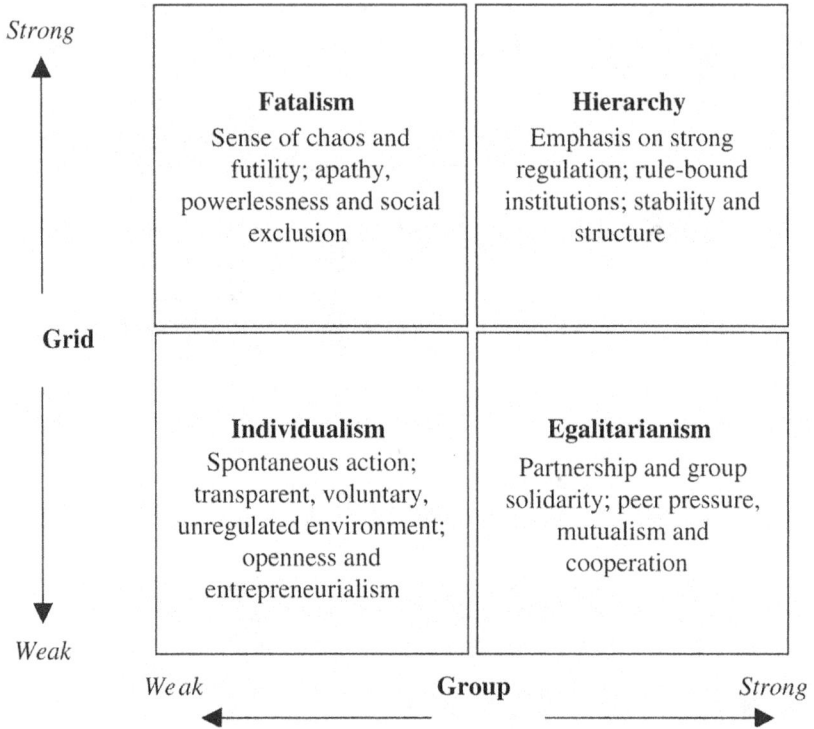

Figure 3.1 The 'group-grid' diagram: diagram of Douglas and Wildavsky's grid/ group typology of world views

Source: Douglas (2007); available free at ResearchGate

Verweij et al. propose each quadrant has its own perspective on human-ecological relationships – hierarchical, egalitarian, individualist, and fatalist – which determines their response to environmental action. You will, undoubtedly, be familiar with the following narratives, all of which can be found within our own culture. To quote:

> In an egalitarian social setting, actors see nature as fragile and intricately interconnected, and man as essentially caring. . . . Voluntary simplicity is the only solution to our environmental problems, with the Precautionary Principle being strictly imposed on those who are tempted not to share the simple life. In a hierarchical setting, actors see the world as controllable. Nature is stable until pushed beyond discoverable limits, and man is deeply flawed but redeemable by firm and long-lasting institutions. . . . In an individualistic setting, actors view nature as resilient – able to recover from any exploitation – and man as inherently self-seeking and atomistic. . . . In a fatalistic setting,

actors find neither rhyme nor reason in nature, and suppose that man is fickle and untrustworthy. . . . "Why bother" is the rational management response.

(Verweij et al., 2006, p. 820)

Each story thus has a different set of heroes, villains, and guidelines for climate change action. Each story also *denies* a particular part of the truth – hierarchicalists deny ecological culpability, individualists deny the fragility and responsiveness of nature, egalitarianists deny the diversity and complexity of human environmentalist attitudes. None of the positions can solve the problem alone. Verweij et al. contend current policy responses have failed because they do not appeal to the multiplicity espoused by all four narratives. The 1997 Kyoto protocol (implemented in 2005) failed because it adopted a largely hierarchical approach – assuming the world's governments would unanimously join forces for the 'public good', strategising and implementing a unified set of environmental policies. Notable exceptions – the governments of the United States, Australia, and Russia – defended their anti-Kyoto stance with individualist arguments, claiming the treaty ignored potential technological advances and future market fixes (ibid., p. 831). Ultimately, the Kyoto agreement failed where future measures could succeed – if they can find creative, sometimes 'clumsy', ways not to alienate significant portions of the global population.

In addition, only one of these four narratives – the egalitarian – recognises the mutually exclusive relationship between people and planet. Our long history of earthly othering and objectification has brought us to a place where we can routinely and subconsciously dismiss our natural embeddedness in the world – so apparently 'logical' and 'rational' reactions to an environmental problem can, in their very practice, completely deny and ignore the truth of our ecological symbiosis.

Everyday denial and emotions

In this final section, I return to individual manifestations of climate change denial; the everyday emotional processing, management, and avoidance we all do. Despite the tragedy of ecological collapse, emotion is routinely left out of public discourse around climate change. Instead, scientific rationalism is used to explain why people are unable to compute the current crisis. Cognition is identified as a primary reason for ecological non-response. Theorists suggest that the problem of individual denial is lack of access to, or understanding of, necessary information. This 'information deficit' model falls short, however, supposing 'information' alone is enough to generate a full and exacting ecological response, and mistakenly assuming that people act rationally on the information given to them (Bulkeley, 2000; Kellstedt, Zahran and Vedlitz, 2008). This approach reflects a general trend of environmental 'emotional denial' – uncomfortable sensations are rendered cognitively inaccessible, requisite of individual management, and socially encouraged to be sidestepped.

Harvard Psychology professor Daniel Gilbert (2006, 2010), shows the human brain is ill equipped to deal with long-term, slow-moving, abstract threats. Designed as a 'get-out-of-the-way' machine, the brain is expert at responding to immediate risks – picking up subtle changes in light, temperature, pressure, sound. We can duck to dodge a bullet or jump out of the way of an oncoming car. Conversely, invisible and incremental differences are largely unnoticed: the slow build-up of impurities in our water and air, remain undetected as hazardous by the brain. Gilbert (2010) notes we respond best to threats which are

1   intentional,
2   immoral,
3   imminent, and
4   instantaneous,

using the example of terrorism as something which triggers all four; it is pre-meditated, violates our moral code, kills instantly, and menacingly appears to be just around the next corner. He says whilst it is easy to enforce protocol against terrorist threats, it is very difficult to get people to do anything for their future selves – from dental flossing, to giving up smoking, to actively reducing their carbon emissions. The brain is simply not wired to respond to threats which we cannot see, perceive, or feel.

Our brain responds most decisively to things we know for certain. Immediate dangers are processed in the amygdala and dominated by the emotional brain, whilst longer-term risks involve intervention from the left-side rational brain. Environmentalist George Marshall (2014) writes we have not yet found a way to effectively engage our emotional brains in climate change – perhaps because we lack explicit and collectively agreed upon concrete examples of its destruction. One of the most successful initiatives over the last couple of years has been the drive against single-use plastic. The force of images and video footage on the effect plastic has in our oceans propelled this issue into the mainstream, as photographs of turtles mutilated by six-pack rings or birds with bottle tops in their stomachs bring the issue into clear focus. A *National Geographic* cover from June 2018 shows an ocean with a giant iceberg stretching up to the sky, whilst beneath the waves it tapers off into two handles – a plastic bag. These images give a visceral response, they make the problem tangible, accountable – something can, and *must* be done. Perhaps if we could see the air or trees around us changing colour we would be permitted to more emotionally fuelled ecological reactions.

Interestingly, however, human emotion is largely omitted from the scholarly discussion of environmental non-response (Norgaard, 2011). Despite intense feelings of anger, fear, helplessness, guilt, and depression, science and policy deny the role of emotions when assessing ecological non-action. In the West, we are routinely encouraged to subdue, ignore, and repress feelings which might 'get in the way' of a rational reaction – falsely assuming we can cut cleanly between logic and feeling. Think again of Extinction Rebellion – a group too emotional, messy, and passionate to be civically credible. Sociologist Arlie Russell Hochs-child (1983) identifies this as 'emotional management' – a means of appeasing and fitting in with social expectations; we must appear level-headed on the sur-face to gain respect, trust, and power in our world. But whilst we are repeatedly called upon to manage our emotions in daily life, we are simultaneously exploited through them by politics and media outlets. Government propaganda relies heav-ily on emotional reasoning to sway voters – Brexit is a crystalline example of how potent emotional rhetoric, to 'defend' and 'protect' against immigration, trumped logical reason – ignoring the material consequences on trade, the market, and international infrastructure. Political structures deny us our feelings whilst also using them as a vehicle for mass control. For climate change, expectations to act rationally set us up for failure – we believe we can respond with logical science and policy, and are surprised when this doesn't work.

Social science shows emotions make a bridge between us and our external reality – feelings are not solely intimate personal experiences but are deeply rooted in and shaped by specific social contexts (Denzin, 1984; Lutz and White, 1986). What emotions people choose to attend to, 'manage', and ignore must be understood within a specific socio-political moment. In his recent book *Meaning and Mel-ancholia* (2018), psychoanalyst Christopher Bollas writes that internal psychic processes mirror what is going on in external society. He argues that the world we currently inhabit is at odds with natural rhythms, laws, and messages. We are experiencing a mass existential disconnect, both psychically and emotionally. An example he gives of the disjuncture between a spiritual reality and its current rep-resentation is 'horizontalism' – the phenomenon whereby 'all ideas are equal and no one thing is intrinsically more important than the other' (ibid., p. 63). We can see this repeatedly on the news, for example, 'where a series of fires in the United States or an impending hurricane, will be given the same air-time as a revolution in Ukraine or a genocide in Africa' (ibid.).

A few weeks ago, I listened as a radio programme segued neatly between reports of a bomb in Mogadishu killing 85 people to news of a pantomime Aladdin proposing on-stage to princess Jasmine. This symbolic levelling of information – from the traumatic to the mundane – reduces demands on us to differentiate between the value of external events, diluting their importance and denying their emotional impact. On a physiological level we can sidestep, or are at least being encouraged to, difficult information and subsequent feelings.

This process of emotional sidestepping is a commonplace management tech-nique for individuals confronting climate change in their daily lives. In her book *Living in Denial: Climate Change, Emotions, and Everyday Life* (2011), Kari Marie Norgaard, a sociologist and specialist in the politics of global warming, provides one of the few Western ethnographies of everyday denial by conducting in-depth interviews and participant observation in the fictionally named 'Byg-daby', a rural community in west Norway. Whilst living in Bygdaby, Norgaard noticed a 'double reality'; the 'collectively constructed sense of normal everyday life' versus 'the troubling knowledge of climate change' (ibid., p. 5). As some of her participants said, 'people want to protect themselves a little bit' (ibid., p. 63) and 'you have the knowledge. But you live in a completely different world' (ibid., p. 3). These experiences reflect both the conceptual limitations of the brain (it does not yet live in a world where climate change poses an intentional, immoral, imminent, and instantaneous threat) and the intersection between individual and socially constructed reality – a world which feels 'completely apart' from environmental danger. Perhaps only when these parameters change – when the risk feels tangible and right in front of us – will we be able to collectively face the ecological truth.

Conclusion

These parameters have already changed; as this chapter goes to press, the world landscape will have transformed beyond recognition. The Covid-19 pandemic has brought the kind of chaos, fear, and uncertainty we expect from climate change – as well as instigating the type of global defence measures and cooperative international policy we will need to protect from it. The pandemic also holds the potential to highlight the reciprocity of human-ecological relations; as human-induced environmental damage is increasingly predicted to affect public health (Singer and Baer, 2009; Pongsiri et al., 2009; Lelieveld et al., 2015; Perera, 2017). Coronavirus – believed to have been contracted from animals – is a type of zoonotic disease, a subsection of infection set to increase as deforestation and ecosystem destruction force animals and people into closer living proximity (Pongsiri et al., 2009; Singer, 2013; Quammen, 2020; Vidal, 2020). This type of 'ecosyndemic' (Singer and Baer, 2009; Singer, 2014) – a term defining the interplay between ecology and human health – forces us to wake up to the reality of our relationship with nature, and hopefully, to take better responsibility for it.

In this chapter, I argued contemporary conceptions of the earth as an 'object' – with no inherent subjectivity or agency of its own – repeatedly casts it outside the realm of human action. We imagine it is a stable, immutable, and permanent mass without any internal ecological processes or dynamics. We pretend human activity exists on top of the earth, rather than inside of – and with consequences for – it. Global capitalism continues to abuse this position; as unrelenting exploitation to facilitate economic growth is endorsed and defended by those benefitting from the status quo – climate change 'denialists' who spend billions in international currency to question, delegitimate, and contradict environmentalist science. Their voice rings loudly through thousands of media publications, broadcasts, and posts which criticise ecological activism as an 'unscientific', 'irresponsible', and 'marginal' concern. The effect is powerful – the *question* of climate change is known to all of us, whether we choose to believe it or not. And how we feel about it – whether these emotions are cognitively accessible, personally managed, or socially manipulated – seems, at present, less important. Whether the emotional wave set in motion by the Coronavirus crisis will be enough to wake us up, and successfully alter the parameters of our current ecological mindset, remains to be seen.

References

Anderson, B. (1983). *Imagined communities: Reflections on the origin and spread of nationalism*. New York, NY: Verso.

Baker, G. (2019). St. Greta spreads the climate gospel. *The Wall Street Journal*, Sept. 20. Available at: www.wsj.com/articles/saint-greta-spreads-the-climate-gospel-11568989306 [Accessed 25 Mar. 2020].

Bancroft, H. and Heale, J. (2019). How extinction rebellion climate change zealots – Including a baronet's Cambridge-educated granddaughter – Are paid £400 a week to bring mayhem to our streets. *MailOnline*, Oct. 12. Available at: www.climatedepot.com/2019/09/20/report-teen-climate-school-skipper-greta-thunberg-is-a-pawn-of-a-consortium-of-greedy-green-venture-capitalists-and-investors/ [Accessed 25 Mar. 2020].

Basso, K. (1996). *Wisdom sits in places: Landscape and language among the western apache*. Albuquerque, NM: University of New Mexico Press.

BBC News. (2019). Extinction rebellion: London protests 'cost Met extra £7.5m'. Available at: www.bbc.co.uk/news/uk-england-london-48269042 [Accessed 25 Mar. 2020].

Begley, S. (2007). The truth about denial. *Newspeak*, 150(7), pp. 20–27.

Bodkin, H. (2019). Parents told not to terrify children over climate change as rising numbers treated for "eco-anxiety". *The Telegraph*, Sept. 15. Available at: www.telegraph.co.uk/news/2019/09/15/parents-told-not-terrify-children-climate-change-rising-numbers/ [Accessed 25 Mar. 2020].

Bollas, C. (2018). *Meaning and melancholia: Life in the age of bewilderment*. London: Routledge.

Bulkeley, H. (2000). Common knowledge? Public understanding of climate change in Newcastle, Australia. *Public Understanding of Science*, 9(3), pp. 313–333.

De Beauvoir, S. (1997). *The second sex*, trans. by H.M. Parshley. Berkshire: Vintage.

Denzin, N.K. (1984). *On understanding emotion*. New York: Jossey-Bass Inc.

Douglas, M. (1966). *Purity and danger*. London: Routledge and Kegan Paul.

———. (2007). *A history of grid and group cultural theory*. Toronto, Canada: University of Toronto Press.

Douglas, M. and Wildavsky, A. (1983). *Risk and culture: An essay on the selection of technological and environmental dangers*. Berkeley, CA: University of California Press.

Dunlap, R.E. and McCright, A.M. (2011). Organised climate change denial. In: J.S. Dryzek, R.B. Norgaard, and D. Schlosberg, eds., *The Oxford handbook of climate change and society*. Oxford: Oxford University Press, pp. 144–160.

Elias, N. (1978). *The civilizing process, vol. 1: The history of manners*, trans. by E. Jephcott. New York, NY: Urizen Books.

Federici, S. (2004). *Caliban and the Witch: Women, the body, and primitive accumulation*. New York, NY: Autonomedia.

———. (2018a). *Witches, witch-hunting, and women*. Oakland, CA: PM Press.

———. (2018b). *Re-enchanting the world: Feminism and the politics of the commons*. Oakland, CA: PM Press.

Foucault, M. (1973). *The birth of the clinic: An archaeology of medical perception*, trans. by A.M. Sheridan. London: Tavistock Publications Ltd.

Freeman, J. (2020). Greta Thunberg and the case of the muddy carbon footprints. *Wall Street Journal*, Mar. 2. Available at: www.wsj.com/articles/greta-thunberg-and-the-case-of-the-muddy-carbon-footprints-11583173445 [Accessed 25 Mar. 2020].

Gilbert, D. (2006). If only gay sex caused global warming. *Los Angeles Times*, July 2. Available at: www.latimes.com/archives/la-xpm-2006-jul-02-op-gilbert2-story.html [Accessed 27 Mar. 2020].

———. (2010). *Harvard thinks big 2010 – Daniel Gilbert – 'Global warming and psychology'*. [video] Available at: https://vimeo.com/10324258 [Accessed 27 Mar. 2020].

Goffman, E. (1963). *Stigma: Notes on the management of spoiled identity*. London: Penguin Books.

Gostoli, Y. (2019). UK climate activists hold 'funeral procession' for the planet. *Al Jazeera News*, Oct. 12. Available at: https://www.aljazeera.com/news/2019/10/uk-climate-activists-hold-funeral-procession-planet-191012183408672.html [Accessed 4 Aug. 2020]

Griffiths, J. (2006). *Wild: An elemental journey*. New York NY: Penguin.

Heidegger, M. (1971). *Poetry, language, thought*. New York: Harper & Row.

Hochschild, A.R. (1983). *The managed heart: Commercialization of human feeling*. Berkeley: University of California Press.

Ingold, T. (2000). *The perception of the environment: Essay's on livelihood, dwelling, and skill*. London: Routledge.

———. (2004). Culture on the ground: The world perceived through the feet. *Journal of Material Culture,* 9(3), pp. 315–340.

Jackson, T. (2009). *Prosperity without growth: Economics for a finite planet*. Bodmin, Cornwall: MPG Books.

Kellstedt, P.M., Zahran, S. and Vedlitz, A. (2008). Personal efficacy, the information environment, and attitudes toward global warming and climate change in the United States. *Risk Analysis*, 28(1), pp. 113–126.

Kuhn, T. (1962). *The structure of scientific revolutions*. Chicago, IL: University of Chicago Press.

Latour, B. (1987). *Science in action: How to follow scientists and engineers through society*. Cambridge, MA: Harvard University Press.

———. (2013). Telling friends from foes at the time of the anthropocene. In: C. Hamilton, C. Bonneuil, and F. Gemenne, eds., *The Anthropocene and the global environment crisis: Rethinking modernity in a new epoch*. London: Routledge, pp. 145–155.

———. (2014). Agency at the time of the Anthropocene. *New Literary History*, 45(1), pp. 1–18.

Latour, B. and Woolgar, S. (1979). *Laboratory life: The construction of scientific facts*. Princeton, NJ: Princeton University Press.

Lawson, D. (2019). Behind science's mask: Extinction rebellion is a doomsday cult. *The Times*, Oct. 12. Available at: www.thetimes.co.uk/article/behind-sciences-mask-extinction-rebellion-is-a-doomsday-cult-8pmbcktqr [Accessed 25 Mar. 2020].

Lelieveld, J., Evans, J.S., Fnais, M., Giannadaki, D. and Pozzer, A. (2015). The contribution of outdoor air pollution sources to premature mortality on a global scale. *Nature*, 525, pp. 367–371.

Levi-Strauss, C. (1966). *The savage mind*. London: Weidenfeld and Nicolson.

Low, S.M. and Lawrence-Zuniga, D. (2003). *Anthropology of space and place: Locating culture*. Oxford: Blackwell Publishing.

Lutz, C. and White, G.M. (1986). The anthropology of emotions. *Annual Review of Anthropology,* 15, pp. 405–436.

Marshall, G. (2014). *Don't even think about it: Why our brains are wired to ignore climate change.* London and New York: Bloomsbury.

Martin, E. (1987). *The woman in the body: A cultural analysis of reproduction.* Boston MA: Beacon Press.

———. (1991). The egg and the sperm: How science has constructed a romance based on stereotypical male-female roles. *Signs,* 16(3), pp. 485–501.

———. (1994). *Flexible bodies: Tracking immunity in American culture – From the days of polio to the age of AIDS.* Boston, MA: Beacon Press.

McGrath, M. (2019). Extinction rebellion: What do they want and is it realistic? BBC News, Apr. 16. Available at: www.bbc.co.uk/news/science-environment-47947775 [Accessed 25 Mar. 2020].

McMahan, B. (2016). Reading the environment. In: L. Manderson, E. Cartwright, and A. Hardon, eds., *The Routledge handbook of medical anthropology.* New York: Routledge, pp. 249–252.

Merchant, C. (1980). *The death of nature: Women, ecology, and the scientific revolution.* New York: HarperCollins.

Monbiot, G. (2007). *Heat: How to stop the planet burning.* London: Penguin.

Norgaard, K.M. (2011). *Living in denial: Climate change, emotions, and everyday life.* Cambridge, MA: The MIT Press.

Oliver, K. (2015). *Earth and world: Philosophy after the Apollo missions.* New York NY: Columbia University Press.

Oliver-Smith, A. (2013). Disaster risk reduction and climate change adaptation: The view from applied anthropology. *Human Organization,* 72(4), pp. 275–282.

Oliver-Smith, A. and Hoffman, S.M., eds. (1999). *The angry earth: Disaster in anthropological perspective.* London: Routledge.

Oreskes, N. and Conway, E.M. (2010). *Merchants of doubt: How a handful of scientists obscured the truth on issues from tobacco smoke to global warming.* London and New York: Bloomsbury Press.

Perera, F.P. (2017). Multiple threats to child health from fossil fuel combustion: Impacts of air pollution and climate change. *Environmental Health Perspectives,* 125(2), pp. 141–148.

Pongsiri, M.J., Roman, J., Ezenwa, V.O., Goldberg, T.L., Koren, H.S., Newbold, S.C., Ostfeld, R.S., Pattanayak, S.K. and Salkeld, D.J. (2009). Biodiversity loss affects global disease ecology. *BioScience,* 59(11), pp. 945–954.

Quammen, D. (2020). We made the coronavirus epidemic. *The New York Times,* Jan. 28. Available at: www.theguardian.com/environment/2020/mar/18/tip-of-the-iceberg-is-our-destruction-of-nature-responsible-for-covid-19-aoe [Accessed 27 Mar. 2020].

Regan, A. (2019). Extinction rebellion: The "reluctant activists" facing criminal records.' BBC News, Dec. 1. Available at: www.bbc.co.uk/news/uk-england-nottingham-shire-50181796 [Accessed 25 Mar. 2020].

Rosenbaum, S. (2006). US health policy in the aftermath of hurricane Katrina. *JAMA,* 295(4), pp. 437–440.

Sabey, R. (2019). Exstinker: Extinction rebellion founder blasted after 11,000-mile flight to central America for luxury break away. *The Sun,* Oct. 14. Available at: www.thesun.co.uk/news/10123037/extinction-rebellion-founder-central-america-luxury/ [Accessed 25 Mar. 2020].

Shapin, S. (1994). *A social history of truth: Civility and science in seventeenth-century England*. Chicago IL: University of Chicago Press.

———. (1996). *The scientific revolution*. Chicago IL: University of Chicago Press.

Shapin, S. and Schaffer, S. (1985). *Leviathan and the air pump: Hobbes, Boyle, and the experimental life*. Princeton NJ: Princeton University Press.

Showalter, M. (2019a). Swedish child climate activist reportedly a tool of Al Gore linked corporate green hucksters. *American Thinker*, Aug. 19. Available at: www.american-thinker.com/blog/2019/08/swedish_child_climate_activist_reportedly_a_tool_of_big_green_corporate_energy_interests.html [Accessed 25 Mar. 2020].

———. (2019b). Sorry, Greta, even your fellow Swedes aren't buying the climate change claptrap anymore. *American Thinker*, Sept. 20. Available at: www.americanthinker.com/blog/2019/09/sorry_greta_even_your_fellow_swedes_arent_buying_the_climate_change_claptrap_anymore.html [Accessed 25 Mar. 2020].

Singer, M. (2013). Respiratory health and ecosyndemics in a time of global warming. *Health Sociology Review*, 22(1), pp. 98–111.

———. (2014). Zoonotic ecosyndemics and multispecies ethnography. *Anthropological Quarterly*, 87(4), pp. 1279–1309.

Singer, M. and Baer, S. (2009). *Global warming and the political ecology of health: Emerging crises and systemic solutions*. London: Routledge.

Turns, A. (2019). I agree wholeheartedly with extinction rebellion, but am I too timid for their fearless tactics? *The Independent*, Oct. 4. Available at: www.independent.co.uk/voices/extinction-rebellion-climate-emergency-crisis-protests-a9142921.html [Accessed 25 Mar. 2020].

Verweij, M., Douglas, M., Ellis, R., Engel, R., Hendriks, F., Lohmann, S., Ney, S., Rayner, S. and Thompson, M. (2006). Clumsy solutions for a complex world: The case of climate change. *Public Administration*, 84(4), pp. 817–843.

Vidal, J. (2020). "Tip of the iceburg": Is our destruction of nature responsible for Covid-19? *The Guardian*, Mar. 18. Available at: www.theguardian.com/environment/2020/mar/18/tip-of-the-iceberg-is-our-destruction-of-nature-responsible-for-covid-19-aoe [Accessed 27 Mar. 2020].

Welsh, T. (2019). We've had quite enough of the law-breaking environmental fanatics of extinction rebellion. *The Telegraph*, June 2. Available at: www.telegraph.co.uk/news/2019/06/02/had-quite-enough-law-breaking-environmental-fanatics-extinction/ [Accessed 25 Mar. 2020].

WHO. (2002). *Environmental health in emergencies and disasters*. Geneva: World Health Organisation.

Zerubavel, E. (2006). *The elephant in the room: Silence and denial in everyday life*. Oxford and New York: Oxford University Press.

Chapter 4

Taking the green agenda out of the margins

Psychological strategies

Andrew Samuels

Here's a summary of the chapter which follows. First, I will link up some aspects of what I call 'therapy thinking' and climate change – with reference to the question of 'mainstreaming' political actions and ideas in this area. Can this be achieved, I ask? Then, in the second section, still in pursuit of the elusive goal of bringing the green agenda out of the margins, I will assert that it's time to praise humanity and human artifice, not to bury them, and pick out some items well deserving of praise. Third, I discuss salient aspects of green politics. In the fourth section, I look at the political desirability and advocacy of sacrifice in the service of the planet by those able to manage it and set this in an economic context. Finally, I will probe what I see as a sort of addiction to apocalypse operating in the West just now. Here's the summary of the summary: first, therapy; second, praise; third, politics; fourth, sacrifice; finally, apocalypse.

Therapy and climate change

Do climate activists ever really want to be in the centre? To be part of the people as a real mass movement? To capture universal and enthusiastic support? In the mordant and suspicious vein of a therapist engaging with what may lie below the surface, I ask: is it possibly the case they cannot be happy anywhere save on the margins? Or, if they do struggle to move into the mainstream, aren't they inevitably going to betray their values and ideals? Or waste their time?

I realise that this is a controversial thing to write about the years 2018–2020. Yet isn't it clear that mainstreaming has become the goal of youth-led movements like Extinction Rebellion? Consider its three goals as listed on its website at the time of writing – winter, 2019 – and note that the word 'government' appears in all of them. What could be more mainstream than to call on the government to do such-and-such a thing?

Government must tell the truth by declaring a climate and ecological emergency, working with other institutions to communicate the urgency for change.

Government must act now to halt biodiversity loss and reduce greenhouse gas emissions to net zero by 2025.

Government must create and be led by the decisions of a Citizens' Assembly on climate and ecological justice.

Is it wise for Extinction Rebellion to be as mainstream as this? Is the government really the key here? It is interesting to review how climate change figured in the 2019 election. Professor Sir John Curtice, who interprets the polls for the BBC, wrote that 'The attention given by all of the parties to the issue of climate change seems to have resulted in an increase in concern about the environment, although it is still relatively low down in the pecking order'.

What about Labour's Green New Deal? My first reflection is that it simply didn't resonate with working class audiences. My impression is this was not because the policy was wrong – but it was presented in an overly detailed and intellectualised way. I mean, who knew what the 'New Deal' was, in the 1930s in the United States? This chapter plunges into both issues: the climate crisis is still relatively low down the political pecking order, and policies are over-detailed and intellectualised.

The link between the climate emergency and our own lives has never seemed clearer. Finally, after decades of activists struggling to push the crisis into the larger consciousness, poll after poll shows public concern, and desire for action, is at an all-time high. The question which became clearer as the year went on was, having achieved what the climate movement always wanted – prominent and positive media coverage, widespread public support, audiences with world leaders – was it possible to effect any actual political change? The spectacle of Greta Thunberg and the larger youth climate movement arriving at international meetings and parliaments and accusing heads of state of hypocrisy to their faces is undoubtedly thrilling. But climate politics itself still seems far from any genuine watershed moment.

Additionally, as the British environmental campaigner George Marshall noted in his critique of Leonardo Di Caprio's environmental film *Before the Flood*, those celebrities and big names warning ever so articulately of the climate change catastrophe which looms are making things worse. Why? According to Marshall, they simply 'ignore entirely the global zeitgeist of popular cynicism about political leaders and institutions' (2015). So: If the facts – 'the truth' – are known by now, why is it proving so difficult to get a majority buy-in for the policies and actions needed? Is there a collective psychological problem? Or is the language and rhetoric being used by climate change campaigners not really working? Answering these questions is what I am struggling with in this chapter. With others, I have been developing what I call 'therapy thinking' in relation to politics for more than 30 years in too many books for comfort. I have pointed out such an activity is truly transpersonal; for politics, like spirit and soul, links people to each other and to everything else is on the planet.

Therapy thinking in the context of climate change has become suspiciously easy. Therapists find it easy to be right when it comes to politics – because one invokes the 'maddening rectitude of the psychotherapist' in which the goal is to prove one's cherished theories – of archetypes, object relations, self-actualisation, whatever – to prove them correct above all else. That is why every single psychoanalytic comment on Donald Trump or Boris Johnson is 100% correct, even when they contradict. It is easy to be right. However, some of the recent history of therapists' engagement with climate change has not been inspiring or reassuring. When I held the elected office of Chair of United Kingdom Council for Psychotherapy, I encouraged the creation of a climate change policy for the organisation as part of its diversity, equality and social responsibility agenda. I can only regret and deplore what seemed to have happened when the proposed climate change policy went to the next board under a new chair. It was said that this is a minority view without sufficient grassroots support. 'What does this have to with psychotherapy?' was asked, and the Board was told that 'Political ideologies have no place in our work'. So, although UKCP recently held a conference on climate crisis, there are many questions left unresolved. For example, if a member takes part in a demonstration and is thereby arrested, do they have to report such a thing to an Ethics Committee? When it comes to politics, sadly and still, it may be a case of 'put not thy faith in therapists'.

This brief opening section is coming to an end. It consisted of some critical comments on the role of therapists and therapy thinking in relation to climate change. The next section makes a positive proposal of what could be done to bring climate activism in from the margins.

It's time to praise human artifice

If we really and truly and seriously want to mainstream ecopsychology and other psychological approaches to climate change, then now may also be the time to praise human achievement and human artifice. On one level, I am thinking of praise, not judgement, for the entire dynamic range of human emotions – the positive ones such as joy, hope and inspiration, and the negative and more difficult ones such as lust, greed and envy. It's impossible to pick and choose; to select only what is nice and appealing. Vitality is not the same as morality, after all.

But it is also time to praise our cities, those achievements of human creativity, aesthetics and social organisation. To praise our squares and piazzas, to praise our restaurants and rejoice in the drinking of alcohol or of coffee, to praise traffic and modern communications. To praise, too, brothels and hospitals, banks and schools. Such celebration has, over time, gone missing from much current environmental discourse. The need to celebrate, to dance and to chant is being more recognised. Yet, this needs to be offset against another tendency, maybe stronger. For I don't think it is helpful to use the language of psychopathology – for example, as George Monbiot often does. Here's an example: 'We need to kick our a*ddiction* to driving' (2016, emphasis added).

Alongside praise of artifice, it is also time to guard against any lingering idealisation of Nature – for this is politically useless and intellectually weak. No-one really knows what 'Nature' means. In his seminal book *Man and the Natural World* (1983), the historian Keith Thomas showed that our present conception of

Nature has a complicated history. But it has a history. Nature changes its nature, so to speak. Thomas sets out the trajectory wherein by around 1800, the world was so irradiated by science, technology and industry that people felt 'begrimed, endarkened and smelly' (ibid., p. 96). So, they sought a sunny, clean and fresh antidote. If they could afford it, they bought country estates. If not, they merely dreamed of pastures and sang hymns about them. This swing to the opposite end of the spectrum – what Heraclitus and Jung called *enantiodromia* – created the modern, romantic notion of Nature. We created Nature!

The snowball of industrialism, Enlightenment and modernity introduced a profound anxiety in European cultural consciousness, to the point of neurosis, over what was being done by civilised humans to the natural world. Between 1500 and 1800, massive doubts emerged over the changes brought about by science and technology in the ways the natural world was perceived. There were many romantic and artistic expressions of this counter-cultural sentiment. Theologians altered their notions about the relations between humanity and the rest of creation to gentle those relations and accommodate a certain decentering of humanity. Naturalists tried to understand and classify other species in non-anthropomorphic terms, thereby respecting their separate existence. Scientists explored links between humans and animals. Moral philosophers urged kindness to animals. In the city, the land came to be regarded as a thing of beauty, fit for contemplation, not only as a useful resource.

In sum, by 1800, people had responded to the anxiety engendered by the brutalising path on which the world seemed embarked. Today's concerns over the limits to economic growth, animal welfare and the fate of the environment may be regarded as descended from these earlier expressions of cultural anxiety. Yet we should temper our admiration for those who could not stomach 'progress'. They did not actually stop its march. Today, animal experimentation and factory-farming coexist with the supreme idealisation of the animal: the child's toy furry animal. As Thomas says, these cuddly creatures 'enshrine the values by which society as a whole cannot afford to live' – an observation he extends to include nature parks and conservation areas (ibid., p. 238).

The revolution in consciousness that Thomas writes about constituted a kind of underground resistance to what was being done to the natural world. This resistance went beyond a reaction to the ruination of nature. The perception of slaves, non-Europeans, children and women also underwent profound changes. As far as women were concerned, the form that liberal anxiety about modernity's denigration of women took was of an oppressive (and convenient) idealisation that restricted women to private and domestic roles. The idealisation of women and the idealisation of nature share similar roots in cultural history in the West: they are both reaction formations, but women and nature remain deeply threatening because the idealisations of them are based on such flimsy and anxiety-ridden foundations.

But by the end of the nineteenth century, we see another swing, this time against Nature. The fight back was led by Nature's great opponent Artifice. My favourite

novel in this direction is *A Rebours* (*Against Nature*), written by the French novelist (and influence on Oscar WIlde) J.-K. Huysmans in 1884. This book, in all its
imaginative perversity and impossible elaboration, is a paean of praise to artifice
and I want to propose Huysman's thoughts like these for us to play with now:

> Nature has had her day; she has finally and utterly exhausted the patience of
> sensitive observers by the revolting uniformity of her landscapes and sky-
> scapes. In fact, there is not a single one of her inventions, deemed so subtle
> and sublime, that human ingenuity cannot manufacture. Does there exist,
> anywhere on this earth, a being conceived in the throes of motherhood who
> is more dazzlingly, more outstandingly beautiful than the two locomotives
> recently put into service on the Northern Railway?
>
> (Huysmans, 1884/1959, p. 125)

I shall conclude this section with an anecdote on the topic of human artifice, which
is what I have chosen to praise. At an ecopsychology conference in Oxford in
2009, I gave a workshop also entitled 'Against Nature'. In it, I distributed sample
phials of many perfumes Selfridges department store kindly gave me. In pairs and
threes, participants used the perfumes, applied them to each other, and compared
notes. It was a smelly exercise and a lot of fun.

Before we did the exercise, I asked who in the audience of around 150 eco-
psychologists wore perfume or its male equivalents. Only one person said she
did. I asked who read fashion magazines in which perfumes are widely advertised.
None, though one person said guiltily that she did this in the dentist's waiting
room.

I then said this showed why environmental activism might possibly fail and
why ecopsychology had truncated itself. For those in the room had, at least as it
seemed to me in the moment, got completely cut off from the role artifice plays
in ordinary human life. Cut off, when you get down to it, from humanity itself.
As far from the mainstream as one can get. I am as frightened of the destruction
of the planet as anyone in the ecopsychology world. But I am also convinced, if
you look in the right way, there is much of value in the fripperies of fashion and
consumerism and it is elitist to deny it. Depth is hidden on the surface.

Green politics: substance and shadow

From a political point of view, my sense is the environmental movement is neither
truly 'for' or secretly 'against' change. It may be both. In the sense that environmentalism represents an opposition to the forms of social organisation established
in the industrially advanced countries during the past two centuries, the environmental movement supports change. But in the sense that some environmentalists
have not caught up in consciousness with the techno-industrial revolutions of the
past 200 years, and may wish to revert to a pre-industrial cultural matrix, environmentalism may be seen as being against the very changes which have already

happened. Maybe, paradoxically, some environmentalism may be regarded as both deeply conservative and wildly radical. The key question, in all its school debating society naivety, remains: Does, or can, human nature change? Oscar Wilde, as mentioned, profoundly influenced by Huysmans, wrote in his tract 'The Soul of Man under Socialism':

> The only thing we know about human nature is that it changes. Change is the one quality we can predict of it. . . . The systems that fail are those that rely on the permanency of human nature, and not on its growth and development.
> (1895/1978, p. 1010)

In this chapter, I have been asking: must the environmental movement fail? I have been suggesting it will fail politically unless the idealisation of nature is somehow moderated. Now I move on to propose further it will fail socially unless it becomes more conscious of certain strands of authoritarianism and depression within it.

A lot of what follows rests on what constitutes 'the real world'. Is the real world the world of the hard-bitten businesspeople who say only by corporate capitalism instituting changes can anything be done about the climate emergency? Or is the real world that of the activists of Extinction Rebellion, which I support and some of whom come to see me as a therapist? For me, the way in which the XR youth conduct their politics reminds of what I was trying to work out in 1993, in *The Political Psyche*, when I wrote the chapter on the 'Political Trickster' (Samuels, 1993, pp. 78–102).

Criticisms in the right-wing press of the authoritarianism of the environmental movement, referring to its 'eco-terrorism' and recent examples of stopping traffic moving in big cities as part of a demonstration, can be frighteningly effective. The latent function of such broad-brush attacks is that entrenched industrial and financial institutions gain succour. As I say, we saw this in the negative reactions in Britain to the rise of a plethora of non-violent and violent direct-action movements, mostly in connection with climate change. Extinction Rebellion, occupying London's bridges and public spaces, is the best known. British readers may recall there were those who cheered when drones – allegedly flown by these 'eco-terrorists' – closed down London Gatwick airport at Christmas 2018, stopping people getting away on holiday. Are these things helpful to the cause of mainstreaming climate change and other planetary issues? Or do they simple demonstrate there is a not-so-hidden authoritarianism in much of the new environmental politics, which could be seen as the latest manifestation of the Enlightenment belief in perfectibility?

Whether this takes the form of equating humanity to the level of fauna (or flora) in a scale of what is valuable or issuing of a whole set of edicts about what is 'good', the tendency is clear to see. Already a backlash is going on. I think that, if it is fair to say that there is environmentalist authoritarianism, stemming from a deeply buried misanthropy, it will be horrendously destructive to the movement. In Jungian terms, this is the shadow of the environmental movement and it would be helpful to become more conscious of it. Then the advantages of the unquench-able human thirst for a better world can be enjoyed – for only things of substance cast a shadow.

Casting an analyst's eye over the information and education material put out by organisations like Greenpeace and Friends of the Earth, I am struck by the one-sided portrait of humanity they present. Certainly, there is much to feel guilty about, much thoughtlessness and destructive behaviour to be owned, much acquiescence in horrid developments to be confessed. But the unremitting litany of humanity's destructiveness may not be the way to spur movement in a more creative direction. The result of too much self-disgust may be the cultivation of a deadening cultural depression which would interfere with environmental action. This is because fantasies of being all bad and all destructive usually lie at the heart of depressive illness. Therefore, environmentalists should try to avoid any pres-entation of ideas about the environment which reflect humanity in an exclusively harsh light. It is not just that guilt-tripping is ineffective politics; my concern is what this one-sidedness obscures.

Going back to Keith Thomas' work, we might also celebrate what careful tend-ing of the earth there has been over millennia. As I have been saying, we might reaffirm the goodness, gentleness and aesthetic sensibility of humanity's artifi-cial, cultural productions – our buildings, cities, art works, and so forth. Politi-cally speaking, it is vital not to represent environmentalism as a concern of the privileged classes or regions, cut off from wider issues of social justice. To begin with, we have already seen the greening of politics is going to be painful, both within Western societies and in terms of the relations between the developed and the undeveloped worlds. A whole host of moral decisions arise when we in the industrially advanced countries call for limits on deforestation in poor countries or advocate their control of their birth rates. We need an educational programme which faces people with these decisions as choices rather than letting choices be made for them by experts who offer protection from the moral implications of what is being done. Otherwise, we end up with a new Western hegemony: we will be more or less OK, but the poor of the earth will be radically worse off as the temperatures and the seas rise.

What is more, we must see through the desirability and efficacy of changes in consumer spending patterns, recycling and veganism to bring about improve-ment. Are we to say that when the going gets tough, the Greens go shopping? If substantive issues of social justice are not addressed, then we *will* just be doing a mere landscaping job.

The question of economic redistribution within advanced societies is going to have to be addressed. If polluters are to pay, prices will rise enormously. The knock-on effects will be dramatic. Many goods we take for granted will be priced out of reach. *I want to suggest that this is a marvellous opportunity!* We are going to have to think about how we live and about how resources are distributed within our more advanced societies – and this will challenge the awesome power struc-tures which exist. The problems confronting the world force a critical engagement with the banks, the multinational corporations, the International Monetary Fund

(IMF) and governments. Without this, poorer people will suffer intolerably. They may – they will – rise up.

Nevertheless, calls for a return to traditional forms of homeworking or the setting up of ersatz agrarian-style communities should be treated with caution. For, in such situations, the lot of women has been and would continue to be an unhappy one. Instead, we should think of greening the cities we already have, making them safer and more pleasant for the groups they oppress – women, children, the elderly. For it has never been demonstrated that agrarian, parochial life is inherently superior to urban, cosmopolitan life. Advocating the tearing down of cities to foster the triumph of nature would be the way of a Khmer Vert.

Younger people will see through any educational campaign idealising nature, leaving out its frightening, harsh and bloody aspects and our ambivalence toward it. Such a campaign would resemble those commissioned portraits of the eighteenth century in which the lady of the manor is pictured dressed up as a milkmaid. The effect was to make nature an acceptable decorative element in the salons of the rich (and see chapter 3 in this volume). We must not do this again.

Politically speaking, the environmental movement still has work to do on a balance between its 'anthropocentric' middle-of-the-roaders and its extreme wing – sometimes called 'ecologism'. Are we doing this for ourselves, for our own benefit and that of our children and other humans? Or is this simply a new gloss on the old exploitative attitude to nature? Should we not be acting for the benefit of an entire planetary organism? Battle lines are even now being drawn up between green extremists and the rest of the community, including 'ordinary' environmentalists. The argument that trees and rivers have rights needs to be assessed so we can distinguish between its potential to inspire action and its gross oversimplifications. Does the HIV virus have 'rights'? Is it ethical to destroy dams or insert into trees spikes that injure loggers?

I hope it's clear that I am not repeating the nostrum, more honoured in the breach than in the observance, that climate changers need to stop telling people they are being very bad boys and girls indeed. Of course, this won't work. But what I am adding is something positive which can be conveyed about aspects of life everyone shares in, to some extent or other. There are some connections between climate change and racial justice that some – but not all – writers overlook and omit. Global warming affects disadvantaged groups in any given society. We saw how, when collective disasters like Hurricane Katrina occur, they disproportionately affect people of colour. And it is in the lands of 'the South' where the effects of our addiction to fossil fuels are most strikingly visible and experienced. With intersectionality in mind, the correlation of race and environment is of huge importance.

Here, it is extremely interesting to note that indigenous communities, such as Native Americans or Aboriginal people in Australia, often lead the fight against carbon and all its derivatives, such as deforestation and open mining. Nevertheless, we should be careful not to Orientalise such people (in the sense of marvelling in a buried, patronising manner). The offensiveness is that we are amazed that 'primitives' can think at all. Assigning 'nature' to them is not a friendly or socially just thing to do, just as was the case when nature was assigned to women.

Allow me to contradict myself. I have been saying that we need to praise cities. True enough. But am I as guilty of an idealisation of the urban as some commentators are of the natural? For, if we have racial and economic justice in mind, then we must admit that cities are not only propitious containers for equality. They could be – but for that to happen, something huge needs to change in urban consciousness. That is the end of an exploration of green politics. Next, we move on to consider the idea of sacrifice in search of the change in urban consciousness I just mentioned.

Sacrificial politics

In this much more depth psychological section, I am in effect linking the psychology of climate change with the whole question of sacrifice. As I have been saying, it is becoming a consensus amongst those who write about climate change and sustainability that the climate crisis and imbalances of wealth under capitalism and globalisation are linked. Economic sacrifices are needed. Because of this consensus, I have been wondering what some ideas about sacrifice might contribute, with climate change and economic justice in mind. We know people will make sacrifices for their children, or for the sake of a cause they believe in, or in the hope of greater benefits in the future (what the economists call 'opportunity costs'). However, sacrifice is a much deeper and wider psychological and historical theme. Sacrifice lies at the heart of the Abrahamic religions (the aborted sacrifice of Isaac in Genesis 22) but is much, much older as a propitiation of the gods. Asceticism has a long cultural history, as does martyrdom.

In Jungian psychology, we talk of the sacrifice of the ego for the flowering of the wider personality in individuation. In art and religion, we contemplate the sacrifice of autonomy and control to something experienced as 'other', whether inside or outside the self. Maybe the time has arrived for psychologically minded people to begin to find an emotional basis for a psychologically considered programme of *economic* sacrifice, calling and naming it as such, rather than waiting for governments to bring it about by fiscal legislation or some other compulsory method – which they are anyway reluctant to do for electoral reasons. I think it is important, if we are thinking of changing the thrust of climate change or any other environmental campaigning, to find a new way of conveying the value, not only the desirability, of sacrifice. Now we come to the promised last section on catastrophe and apocalypse.

Addiction to apocalypse

I want to discuss why, when it comes to climate change, it is still quite often a case of 'Eat, drink and be merry, for tomorrow we die'. I want to give my own suggestion as to why there is the denial, disavowal and despair so many climate change psychologists write about in such interesting ways. Yes, what follows is exaggerated – but as Theodore Adorno (2006) wrote: 'In psychoanalysis nothing is true except the exaggerations'. Exaggeration is at the heart of satire, and much research shows the usefulness of a Trickster element when it comes to communicating scientific information. Here's an example: 'We do not care about planet Earth', four French scientists declared in February in the journal *Trends in Ecology and Evolution*. If humans are exhausting the planet's resources, they wrote, it's earth that needs to adapt – not us. The authors issued a warning: 'Should planet Earth stick with its hardline ideological stance . . . we will seek a second planet' (Quoted in Preston, 2018, no page given). This got a lot of attention.

The use of satire to defuse anxiety-driven rejection of the already existing planetary catastrophe is a contribution to what I called in *The Political Psyche* (1993, pp. 78–102) the 'political Trickster'. The politics of Trickster are often still overlooked. We perform our environmental politics within a fantasy of our own seriousness. For the Greeks, the arch-Trickster was Hermes, with his tendency to play jokes, to lie, to cheat, to steal, to deny reality and to engage in grandiose fantasy.

Genuine Tricksters, from Coyote in North America to Ananse or Eshu in West Africa, follow this pattern, undermining the prevailing organisation of power and even the perceived structure of reality itself. Tricksters can certainly personifications of primary process activity, challenging and disregarding the laws of time, space and place. Rather than judge and condemn, let us speculate about why Trickster mounts this challenge. He does it *precisely to test the limits of those laws*, the bounds of their applicability, and, hence, the possibility of altering them. At the moment we say this, the political referents of the Trickster are revealed. He enters the arena of climate change and environmental despoliation. Challenging the limits of laws, their applicability and the possibility of altering them – and doing this in an ideological climate which is hostile to such a challenge – is *the* classic progressive political project.

Claiming the Trickster for environmental politics might seem like the most crass over-interpretation. But if Trickster's political theorising can retain his own capacity for shock and irony, then no more than a little damage will have been done to him. If involving the Trickster in political discourse does not injure the Trickster, then what does it do to our conception of politics, and environmental politics in particular? There is a conventionally moralistic nature of most depth psychological analysis of the fate of the planet. For example, the home page of the estimable Climate Psychology Alliance website carries the slogan 'Facing Difficult Truths'. Well, what could be more conventional than to advocate the facing of difficult truths? The language is so middle-aged, middle-brow, middle class.

From this perspective, driven as it happens by psychoanalysis, Trickster's mendaciousness and self-deception are misunderstood as immature, obscuring his transformative and generative aspects. We need to be both sensible and other-than-sensible. Green politics requires a ceaseless dynamic between a passionately expressed, codified, legally sanctioned set of principles and certitudes (original morality) – and a more open, flexible, improvised, tolerant morality which is basically code-free (moral imagination). These two aspects of political moral-ity are present in varying degrees in any political problematic. It is important to resist the temptation to see one of them as somehow more advanced, rising from the ashes of the other. Certitude and improvisation are *equally* valuable and, even assessed from a conventional psychodynamic perspective, they are equally mature.

It is easy to see that a political morality based exclusively on improvisation would be too slippery by far and would contribute to a culture in which anything goes. But a political morality based exclusively on principle, law and certitude would be equally problematic (in psychological language, equally 'primitive'). To begin with, laws are not politically effective on their own; legal codes reflect and depend on the distribution of wealth and power. Moreover, political princi-ple easily becomes ossified and used to gain control over others. Finally, codi-fied political and moral prohibitions do not always work, as the prevalence of theft or adultery demonstrates. I accept that Trickster's discourse may seem like garbage to some readers. Yet his refusal to say definitively *this* is the only real-ity (for example, economic growth, fracking, coal mining and industrialisation) and *this* is Utopian fantasy (for example, radical reform or even revolution in planetary politics) is in itself a profound, political statement. Viewed this way, Tricksters contribute to the abolition of every kind of exploitation or oppression, be it directed against a class, a party, a sex, a race – or a planet.

I now turn to 'Apocalypticism' – the belief there will assuredly be an apoca-lypse. When deployed by Puritan sects, the term apocalypse originally referred to a revelation of God's will, but now usually refers to the belief the world will come to an end very soon, even within one's own lifetime. This belief is usually accompanied by the idea that civilisation will soon come to a tumultuous end due to some sort of catastrophic global event. The notion the world is coming to an end is fairly called 'archetypal', found in many religions, paths and 'ways'. This is what gives apocalypse the power to possess groups and individuals who do not belong to Puritan sects. To possess all of us, perhaps? Is this what has happened in relation to climate change? If so, then we have the beginnings of a theory as to why so many people in the Western countries have so little interest in the matters we are discussing today. They prefer not to.

Climate change and planetary degradation inspire images of an apocalypse which one would imagine to be horrid, but which may be oddly pleasing and reassuring. The breakdown will happen, nothing to be done about it. And that could be for some people an oddly reassuring thought – because it justifies inac-tion. Fantasies of an apocalyptic end are rooted in reality, and it is right to point them out. But these may be deep signs of a self-punishing contempt for ourselves. Apocalypticism is not based on fear of an end, but rather on desire of it.

Perhaps some people think we deserve to perish like this. Is this perhaps why many people don't talk about climate change?

Perhaps this self-loathing is a shadow element for many people, including me, with concerns about climate change. It exists alongside our excitement at witnessing the rise of a responsible tending for the planet, and the flowering of depth psychological interpretation of climate change denial, disavowal and despair. We desire – we actually want – the whole terrain temple to crash down. It is a tad exciting, a macabre spectator sport, a form of political pornography – masochism in an environmental setting. Why do I end my chapter on this note? Because I feel obliged to say, in inflated and prophetic mode, it is *the very love of catastrophe* that contributes to our paralysis. What do we want? Apocalypse. When do we want it? *Now*. We climate change campaigners can never move to the centre if we don't think about this thing of darkness that is holding us back. Hamlet got it:

> To die: – to sleep;
> No more; and, by a sleep to say we end
> The heart-ache and the thousand natural shocks
> That flesh is heir to, 'tis a consummation
> Devoutly to be wish'd
> (Act 3, Scene 1, lines 21–6)

References

Adorno, T. (2006). *Minima moralia*, trans. by E.N. Jephcott. London: Verso.

Huysmans, J.K. (1884/1959). *Against nature*, trans. by R. Baldick. Harmonsdworth: Penguin.

Marshall, G. (2015). Engage centre-right voters to put climate change on the political platform. *The Guardian*, Apr. 15.

Monbiot, G. (2016). Our roads are choked. We're on the verge of Carmageddon. *The Guardian*, Sept. 21.

Preston, E. (2018). Using satire in science communication. *Undark: Truth, Beauty and Science* (no page).

Samuels, A. (1993). *The political psyche*. London and New York: Routledge.

Thomas, K. (1983). *Man and the natural world: Changing attitudes in England 1500–1800*. London: Allen Lane.

Wilde, O. (1895/1978). The soul of man under socialism. In: *Complete works of Oscar Wilde*. London: London Book Club.

The moons of Jupiter

Those who refused to look through the telescope
of seductive Galileo were perhaps far-sighted:
I think they foresaw the rise of the lightning conductor,
the improvements in surgical hygiene, the increase in the production of
    pig-iron,
the scooting-off-down-the-track of the first steam-engine, the formulation of
    Coca – Cola,
the launch of Sputnik, the worldwide sale of the dish-washer,
the ascent of the share-price of Apple, the arrival of universal surveillance,
the poisoning of all the world's oceans, nuclear weapons in the hands of
    madmen,
the choking of the earth's atmosphere, the dying of all the world's
    animals . . .
and they said: 'You know what, Galileo Galilei? –
we are far-sighted; we have telescopes of our own; we
have seen the moons of Jupiter. Thank you! Good-bye!'

D.M. Black

from *The Arrow Maker* (2017, p. 24) London: Arc publishing

Part II

Images and imagination

Chapter 5

Cultural crisis

A loss of soul

Chris Robertson

> The dread and resistance which every natural human being experiences when it comes to delving too deeply into him or herself is, at bottom, the fear of the journey to Hades.
>
> Jung, C.G. (CW 12, para. 439)

> Most people guard and keep; they suppose that it is they themselves and what they identify with themselves that they are guarding and keeping, whereas what they are actually guarding and keeping is their system of reality and what they assume themselves to be. One can give nothing whatever without giving oneself – that is to say, risking oneself.
>
> James Baldwin (1963, p. 100)

There is a pervasive malady in the air. Terrorist attacks, violent social upheavals, species extinctions and natural catastrophes like tsunamis leave us with an uncanny sense of menace. The ground is wobbling; sometimes with earthquakes but more usually with culture quakes – ruptures in our social expectations which leave us uncertain and anxious. In the quotation opening this chapter, Jung describes his own experience of entering a dark time in his life. A descent into depths involves passing through the dread he speaks of, being willing to enter the unseen, below the surface. Such descents happen at formative thresholds when a person, group or collective reaches within to find the resources necessary to meet a crisis.

In our present climate emergency, Jem Bendell (2018), professor and founder of the Institute for Leadership and Sustainability at the University of Cumbria, has outlined the reasons for a likely social collapse of our tightly enmeshed economic and distribution networks. The work of Extinction Rebellion and the school strikes bring home the real possibility of the end of the human species, yet most of us continue with business as usual. Governments declare climate emergencies but fail to act. The rates for suicides and self-harm in young people are reaching epidemic proportions. Something is happening but we don't want to know what it is.

As Jared Diamond, author, anthropologist and historian, evidenced in his book *Collapse* (2006), historically cultures are often blind to their own degrading and collapse; the cultural ego does not want to sign its own death warrant.

Earth system science predicts several tipping points in the complex ecology of our planet including ice sheet disintegration, ocean acidification, carbon dioxide induced temperature rise, biodiversity loss and rainforest die-back. These anthropogenic, human-induced changes affect the whole planet. Humans have tipped the earth out of its stable Holocene equilibrium. Why is there such disavowal of this crisis? What really needs tipping is our own thought.

Despite this extraordinary denial on the level of public discourse, our collective unconscious has got the message. Yes, we are in grave danger, we feel a collective spasm of existential dread for our species. Yet our cultural ego has stopped listening, basking instead in technological commodities and the fetish pleasures of selfies. Psychological services are inundated with requests for help with 'eco-anxiety' (see Costing the Earth, 2019). Some blame the apocalyptic media stories and warn the media not to upset people, but depression, helplessness and panic attacks occur in children who do not follow the media. It is more accurate to call this distress an outbreak of environmental awareness rather than add it to the lexicon of medical terminology. The young generation can see that the 'emperor has no clothes'. They cannot wait for the old belief patterns to die out. The school strikers know we do not have time.

I suggest these collective symptoms are a cascade of traumas humans suffered in our evolutionary journey from hunter-gatherers which include: the slave trade, the industrial revolution, colonisation, the Holocaust and atomic bombing. Residues of past trauma mixed with traumatic activation in the present distorts what is perceived as threatening leaving us in a perpetual state of vigilance. The result of these unattended traumatic stresses shows itself in cultural fatigue – dissociative, depressed symptoms defended against by a manic drive for comfort, success and celebrity.

As American author and journalist David Wallace-Wells (2019) wrote in *The Uninhabitable Earth*, they are written clearly enough in floods, storms, tsunamis, droughts and plagues. Climate change is a natural force outside of human control, a challenge to human exceptionalism and superiority. Catastrophic fantasies of ecological collapse and ecocide cannot be explained away. They are in our dreams and have penetrated the membrane of our collective ego. We know from chaos theory tipping points occur at thresholds between stable patterns. As early climate modeller, Edward Lorenz (1993, p. 134) explained with the metaphor of the 'Butterfly Effect', existing systems are vulnerable to small shifts which amplify into large changes. At these vulnerable thresholds, potentials for transformation are at their most powerful. Naomi Klein, author and social activist, points out in *The Shock Doctrine* (2008) how the political right use these moments of vulnerability to re-impose more forcefully the restrictions they want. They are also portals to previously unthought possibilities.

Our culture is at a cultural tipping point. Do we see the signs or is it too scary to face? Can our risk-averse culture be turned away from a flight into its escapist comforts? Like any existential crisis, our present ecological emergency offers opportunities to re-think what it means to be human. Defining our species through cultural complexes such as *progress*, *superiority* and *dominion over nature* needs to end, if we are to survive. We are not superordinate beings but one species among many. Such a radical adaptation might include:

- relinquishing our cherished special position of dominion
- acknowledging existential shame for betraying our instinctual inter-dependence
- taking responsibility for our violence and hate towards other beings
- restoring our sense of earth's sentience and repairing damaged relations with the other-than-human
- reclaiming our capacity to give birth to a fresh imagination.

This radical adaptation requires both letting go *of* social defences which blanket our ability to recognise the crisis and letting go *into* a revitalised imagination. In the next few paragraphs, I will focus on the relinquishing of what we have become unconsciously attached. If the psychological and often painful work of letting go is inadequately engaged, then the imaginal work of opening to unknown possibilities is in danger of being escapist.

Constraints around letting go of social-cultural defences relate directly to social anxiety. Menzies-Lyth (1970), a founder of the Tavistock Institute and pioneering social researcher, explained the dysfunction in the nursing system of a general hospital as a defence against anxieties aroused by exposure to the painful suffering of patients. Unconscious defences against such pervasive anxiety were social constructed through rituals and routines of nursing practice. Because these social defences were operating unconsciously and deeply built into the nursing system, they were difficult to change. As James Baldwin, the author, playwright and Black activist quoted earlier, says (1963, p. 100), 'what they are actually guarding and keeping is their system of reality and what they assume themselves to be'.

Although the concept of social defences was developed within organisational systems, it has clear relevance to the wider social-cultural systems, especially in transitional times when rapid change creates anxiety. When such stresses are combined with the loss of community and a cultural complex of hyper-individualism, no one can be trusted. Narcissistic defences to avoid vulnerability are now a cultural phenomenon. The felt necessity in a narcissistic culture which worships celebrity is to maintain the illusion of self-sufficiency – the showbiz grin saying, 'everything is well with me'. The fault line exposing vulnerability or becoming 'a loser' is avoided by constant hyper-vigilance in this costly defence.

We live in a hollowed-out culture which has lost its soul and stands at a threshold, a cultural Kairos as Carl Jung suggested in *Modern Man in Search of a Soul* (2001). When nothing has intrinsic value but only an instrumental use for potential exploitation, we stop seeing others as beings in their own right. We stop meeting others without functional aims. What started with a de-souling of outer nature is now eating its way through our culture. This hollowing-out works through commercialisation and commodification of personal intimacies and tragedies in Reality TV and social media.

Breakfast talk with Pat and George: an imaginal episode

The longing for and terror of this something 'other' can be played out in domestic dramas:

Pat:	I'm dying.
George:	*(reading the paper, not responding.)*
Pat:	I'm dying.
George:	You're what, dear?
Pat:	I'm dying.
George:	You're dying to what? To go shopping?
Pat (a bitter tone):	No, I am dying.
George:	*(a little concerned, looking up.)* Have you been to the doctor?
Pat:	I can't get an appointment and anyway, there's no cure.
George:	Did you try calling a help line?
Pat:	Yes.
George:	What did they say?
Pat:	That it is environmental melancholic solastalgia.
George:	*(worried, confused.)* Is that bad?
Pat:	Yes, it's fatal. Death is the only cure.
George:	*(nonplussed.)* Is it passed on through personal contact?
Pat:	No, it passes through thought.
George:	Through thought?
Pat:	Yes, even through the news.
George:	Surely not.
Pat:	Yes, there is no protection. Did you read about the threats to biodiversity? It is heart-breaking. I feel sick to death we carry on as normal.
George:	*(silent, crestfallen.)*
Pat:	Don't worry, George. You're immune.
George:	Maybe we ought to go shopping.

We don't know what precipitated Pat's crisis. Perhaps she has an underlying disaffection with her consumer lifestyle, or an old emptiness resurfaced after her children have left home. It would be wrong to assign her crisis solely to her personal history, though it may make her susceptible. We could guess she is suffering from an acute sense of climate distress, sometimes dubiously labelled 'eco-anxiety', about the state of the planet and our human destructiveness. Her 'sickness' is not a medical condition requiring tranquilising, but a heartfelt despair. Thinking less literally about her suffering, when Pat says she is dying, it is a sense of being alive, of meaning, of connection which she is missing. This loss of vitality is a loss of soul.

Pat mentions a diagnosis of solastalgia. This term, coined by the Australian philosopher Glenn Albrecht et al. (2007), gives a name to a homesickness caused by environmental change: an endemic sense of place has been violated. We may be afflicted by solastalgia as distress about the loss of other species, our kin, the impact of any environmental damage to our home environment and the irreversible losses to our home planet.

Pat grieves the loss of the world she knew in the present sixth great extinction. Plants, trees, animals and humans are dying along with cultural traditions, rituals, languages and skills lost, possibly forever. Loss of soul connection, of reciprocal intimacy, of Eros, leads to the parched dryness of an unquenched thirst. Ensouling moistens, softens and lubricates, giving us a sense of depth. Without it, we are arid; superficial, a sense of meaning in life elusive.

The day world, the world of ego functioning and control, is breaking apart. The underworld, the feared Hades of which Jung spoke, appears as we 'go under'. Relinquishing ego consciousness does not happen easily: loss of control feels like dying. We need different, even perverse sensibility to engage. Why perverse? Because fantasies which allow entry into this underworld are not the healthy sunshine ones of the dominant ego. What we may pathologise as 'morbidity' may be fascination with death within a culture avoiding it. The refusal to go under, to maintain the ego's earthly plans for success, lead to an ecocide as preferable to ego loss. But, like green shoots finding their way through concrete pavement, the unconscious operates through portals of happenstance, synchronicity and dreams.

Pat and George: imaginal episode 2

George had a disturbing dream. He hesitates to expose it.

Pat: What is it, George?
George: What is what dear?
Pat: You look like you have been stung or that something is eating you.
George: Well yes: probably both . . . *(reluctantly)* I've had a dream, and it won't leave me alone.
Pat: Would you like to share it with me?
George: Not really, but it feels like I have to; I am compelled to tell you.
Pat: Go on, then. I'm listening.
George: I am somewhere in the city and it is getting dark. I join people who seem to be processing somewhere. I cannot see them but feel drawn to join in. After a while, we reach an open space by large buildings and, glowing in the moonlight, appear strange beings. I am fearful but fascinated. What are they?

    The leader of our procession slowly reaches out his hand to touch or not quite touch one. He beckons us to copy. I am in a frenzy of fear and excitement. As if in a trance, I move towards one of the glowing shapes. As I reach out my hand, I feel a powerful cold. I realise these are blocks of ice emitting an eerie blue light.

Pat: That was a powerful dream. Are you OK? You look a bit white.

George: I could do with a cup of tea.

Pat: I'll make you one. What a coincidence. I was just reading about an art installation outside the Tate Modern made of blocks of glacial ice brought from Greenland by a Scandinavian artist. He wants us to connect with the melting Artic, but you connected without even going there. I was wrong about you, George; you are not immune.

In *The Plague*, the social pestilence Camus describes, 'everyone has it inside himself, this plague, because no one in this world, no one, is immune' (1960, p. 229). Here, the psychic infection woke George from his entrancement with the mundane and safe. He is touched. The dream penetrated his estrangement. It awakens him.

Sometimes what threatens our known ego-contained world, also fascinates us and draws us to it. As Andrew Samuels (Chapter 4 of this volume) says of apocalypse, 'Apocalypticism is not based on fear of an end, but rather on desire of it.' There is certainly a fascination with doom, horror and tragedy showing the shadow side of our secure, safety-first immunised culture. The abject, like the cast-off icebergs, emits a perverse pulse of forbidden pleasure. Like the dark sun in alchemy, the apocalypse can act as a strange attractor.

Chaos theory brought us the powerful metaphor of *Strange Attractors* to explain how an apparently unordered system has recourse to patterns not readily recognisable or familiar yet not random and sometimes profoundly beautiful. Transferring the metaphor of strange attractor to social psychology, allows for the possibility of deep connections within the collective unconscious. Jung (CW 8, para. 818–968) coined the term synchronicity to speak of meaningful acausal connections, which are not explainable. Once we move into this way of thinking, we start to view *order* as having potentialities and *chaos* as having sensitivities.

Even while we avoid it, the dark pulls us toward it. The breaking open of our ordered lives may involve descent into darkness. The descent into apparent dark chaos typically entails destructiveness or violence. This might be self-harm, suicidal ideation and dissociation. There is something to be feared and it is our selves. Giving a place to these dark and troubled forces, strangely (at least to the ego) evokes soul.

Michael Ortiz Hill, author and practitioner of traditional African medicine, perceives a similar threshold in his fascinating book *Dreaming the End of the World* (2005). Through studying hundreds of apocalyptic dreams, he suggested that to transform a literal apocalypse into an initiation, we need first to enter deeply into archetypal fears. In knowing these fears, they no longer bind us to act out unconscious reactions. Initiatory connection can ensue. In the face of the numinous, Hill writes our soul is stripped bare. 'It suffers the raw truth of the moment, its conundrums and heartbreak, and witnesses the death and rebirth of the self/planet' (ibid., p. xix).

Hill reminds us *apokalypsis* in Greek means revelation or the tearing away a concealing veil (ibid., p. 22). He suggests apocalyptic dreams give a 'vivid and raw picture' of ourselves against the destructive realities of present times (ibid., p.50). We can also reflect on the alienated relationship grown up between an urbanised human realm and the wildness of nature. This separateness, so intrinsic to the Promethean venture of controlling external nature, banishes soul as the mediating container between our cultural ego and wild nature.

Facing into dream images of annihilation, environmental disaster and ecological collapse can be terrifying and liberating. The power which fuels destructive rape of our earth can find a place in our collective psyche if its destructive reality is faced. Jonathan Lear, professor of Philosophy and Social Thought at the University of Chicago, in his book *Radical Hope* (2006) drew on the story of Crow Chief Plenty Coups to imagine the resources and ethical values needed for the Crow nation to adapt to a new way of life after their traditional ways had collapsed (see also Anderson and Robertson, 2016).

For the Crow to adapt, they needed a challenging reversal of normal values. Similarly, when we talk about the end of the world, our capacity to make sense of this is very limited. Even if we say it is the end of the world, *as we have known it*, meaning the collapse of the social, economic and cultural norms, this seems circular. The only world our socialised mind has known *is* this world – as we know it. It gets far more peculiar when we start to imagine a world after our species is extinct. Who would be observing, witnessing this dissolution? Is there a sentience beyond human consciousness? Might there be multiple participants in co-creating the biosphere?

There is something mysterious about our language and conjectures when we explore unknown worlds. Timothy Morton, author and philosopher, admired by the singer Björk for his thought on species inter-dependence, explores this perplexity in *Being Ecological* (2018). He suggests we are entangled in the web of language and struggle to describe appearance or experience. And this very struggle to face into external reality creates another twist entwining us further in the web. A key to the difficulty thinking about global catastrophe, Morton suggests, is the phenomenon of the *Hyperobject* which defies our perception of time and space because its duration enormously exceeds our own lifespan (and see chapter 11 in this volume). Climate change is just such an object.

For instance, with the notion of the Anthropocene: is this a geological reality we are transiting or is it a concept within our minds, which we could exchange with notions such as the Capitalocene? From a geological perspective of a supposedly inert planet, Western culture is likely to leave for posterity sediment composed of concrete and plastic. But what of the live, animate planet? Despite being regarded as dead, Gaia is making impacts into the human world. The geophysical is intruding into the geopolitical and not as a nourishing mother figure. The dark side of Gaia has always been there but to the extent she was recognised, many humans have preferred the idealised version (as with the projection onto mother figures). This coming to terms with Gaia as a force beyond our control denotes a threshold for the collective ego.

The Anthropocene marks the apotheosis of the human species and a pending collapse threatening poverty, destitution, starvation and drought. It creates a tear in the social fabric; felt as a cultural trauma with frightening loss of stability. We may need a psychology of collapse such as explored by Jared Diamond (2006). This collapse has been the focus of the Dark Mountain project (https://dark-mountain. net) whose poetic and imaginative writing face into this dark reality tracing the deep cultural roots of the mess the world is in. More recently, American novelist and journalist Jonathan Franzen has asked (2019), 'What if we stopped pretending? The climate apocalypse is coming. To prepare for it, we need to admit that we can't prevent it.' While the refusal to take the opiate of false hope is courageous, an apocalyptic focus on collapse is too binary – either we stay in denial or we face collapse.

It seems to me vital to move our narrative away from the literal and scientific reality. Even if climate science provides evidence which is increasingly difficult to ignore, we need something different to take us over the bridge of our cultural crisis, away from notions of a dead earth into an alive one. The objectification of *nature* as something 'other', of which we can be outside observers, has opened the gate to destructive exploitation. Technical science, after all, has been the hand-maiden of the Beast – the industrial complex relentlessly consuming everything in its path.

Jung (2009) in *The Red Book*, his extraordinary account of his descent into an unconscious of previous unfaced terror, modelled a third possibility between denial of a cultural crisis and social collapse: a night sea journey into the depth of soul which reverses the cultural momentum of ego colonisation. It de-literalises the apocalyptic fantasy of outer destruction into a journey into Hades. Descent into those dark saturnine feelings involves a process of acceptance, falling apart to open pathways of radical transformation. These are imaginal pathways through the liminal territory of the transit to the Anthropocene. They are not rational planning routes but, like dreams, offer glimpse of new realities unseen by the dominant anthropos.

Approaching the Anthropocene as a dream

In this section, I attempt to free us from the literal constraints of the day world and enter the imaginal other world we visit in dreams. Imagine the rational technical mind has itself been a strange dream; a dream of power over, dominion, freedom to control, comfort and material riches. This dream of progress, faltering with the broken promise that our children will inherit a better life, is sometimes called the 'American Dream'. This dream is now losing its power, its capacity to entrance. Despite this, the fantasy of a techno-future where the mind is finally freed from the limitations of the human body, continues in thrall to the dream of progress. Robert Romanyshyn, author and professor of psychology from Pacifica Graduate Institute, writes of the destructive power of this in his *Frankenstein Prophecies* (2019), the gradual unravelling of the omnipotent fantasy of power over nature (and see Chapter 6 of this volume).

Deep ecologist and author Thomas Berry (1988) suggests we need an *earth dream* as a counter entrancement – opposing the entrancement of the industrial complex. To imagine an earth dream, we need to disengage from the cultural of hyper-individualism. Our dreams are not the property of our egos. Some dreams may be neurotic reflections or wish fulfilments, but others are not *ours*. They visit us. We are in the dream. So, what might be this earth dream that is dreaming us?

James Lovelock, inventor and originator of the Gaia hypothesis of our planet's self-regulating sentience, could be considered an earth dreamer. In recognising the earth system as self-regulating, he attributes some agency to Gaia, meeting scepticism from other earth scientists for whom a cybernetic homeostasis is the paradigm limit. He took a step further along this road in attributing a capacity to fight back against destructive consumption in his book (2006), *The Revenge of Gaia*. The revenge is metaphoric, like the vengeful Greek Furies, but the emotion of revenge is problematic within earth systems.

For all his inspiration, Lovelock seems held back from fully entering an earth dream. Shifting to a description of Gaia, David Abram (2010, p. 8), an American ecosystemic thinker, author and cultural ecologist, writes *In the Depths of a Breathing Planet*:

> The world we inhabit is not, in this sense, a determinate or determinable set of objective processes. It is flesh, a densely intertwined and improvisational tissue of experience. It is a sensitive sphere suspended in the solar wind, a round field of sentience sustained by the relationships between the myriad lives, the myriad *sensibilities* that compose it. We come to know more of this sphere not by standing apart from our bodily experience but by inhabiting our felt experience all the more richly and wakefully, feeling our way into deeper contact with other experiencing bodies, and thus with the wild, inter-corporeal life of the world itself.

From the perspective of a sentient breathing planet, it is problematic to attribute 'revenge' to the myriad lives of this wild, inter-corporeal world. The separating out necessary for the sense of grievance and its revenge is a particularly human perspective. There may be self-regulating negative feedback in the earth system to human excess exemplified in the climate emergency, but it is not retribution. We may feel guilty about our complicity in the destruction of our home planet and like children imagine the mother's, or God's, angry retribution. James Cameron's film *Avatar* (2009) dramatised this regulatory tension. Initially the narrative had 'nature' as neutral to whatever happens, she would not take sides, but then the dangerous animals are activated to support the indigenous and attack the human predators. Gaia is not neutral; she is a self-regulating agency which will attempt to balance earth systems, even if the not-so-special human species are damaged in the process.

If, following Abram, we shift from distance to inhabiting a felt experience of intimate connection with life on earth, we could imagine being invited by Gaia into her dream. An invitation to a different collective dream beckoning us to counter the spell of the technological progress, then we could call it a Gaian dream: a dream we are party to, but is so much more than our part; a dream which includes and enfolds us, provided we can decentre from our self-importance.

Such dreams do not wait to be invited but makes themselves present despite our preoccupations. Well before the start of Re-Vision, the psychotherapy training centre I co-founded in 1988, my partner and I had a series of conjoining dreams. One night, one of us would have the dream, the other continued the dream the following night, and so it went on. Something *other* was coming through us. We might say we were *being dreamt*. We were in the grip of a shared dream which held a *telos* – a sense of being touched by fate. A much later analysis of the dreams guided us to the formation of Re-Vision (see *Transformation in Troubled Times*, 2018).

This inter-connection of dreams is explored in social dreaming, a dreaming matrix whereby participants share dreams not for individual meaning or interpretation but for their value to the community of the dreamers. Gordon Lawrence, social scientist at the Tavistock Institute and founder of the field of social dreaming, saw the possibility of dreaming socially, not about 'me' but about the human condition. His book *Social Dreaming at Work* (1998) sets out some of the principles. A wide application of social dreaming to diverse fields has followed his pioneering start. This includes the social dreaming practitioners and authors Manley, Gosling and Patman (2015), who explore the ontological status of such dreaming in *Full of Dreams: Social Dreaming as Liminal Psychic Space*. They define liminal as experience and knowledge that is 'unthought' and in various stages of transition between what Bion (1962) called Beta and Alpha functions, somewhere in between or fluctuating between unconsciousness and consciousness; where the shared associative imagination blends with affect as experienced in images, symbols, signs and representations – in other words, a zone of dream images and their associations.

They distinguish their use of the notion of liminality from that of Victor Turner (1967), the anthropologist considered to have rediscovered the importance of liminality, which concerned itself with states of identity and social transition. They cite psychoanalyst Donald Winnicott's ([1971] 1991, pp. 41, 47, 53) use of 'potential space' as an in-between space which fosters play and as-if fantasy. While liminality can be seen in developmental phases such as adolescence, it is more than a developmental transition. It can also represent a threshold between worlds where, because nothing is certain, transformation becomes possible.

Psychology of borders

What these experiences of liminal space evoke is a psychology that explores the marginal edges of experience, a space between worlds. At junction points and thresholds, normal laws fail to operate. We enter a liminal space where the old means of regulation do not work, but we have not yet developed new ones. These borderlands lie, like their territorial counterparts (see Farley, and Roberts who describe in *Edgelands* (2011) journeys into England's true wilderness), at an in-between reality not governed either by what has come before or what is attempting to manifest. They are intrinsically uncertain places to be in which small changes can be amplified.

While being chaotic in nature, they are a source of creative potency. Winnicott ([1971] 1991) considered such transitional spaces to be the location of potential cultural transformation and provided continuity for our species which transcended individual existence. The mysterious, chaotic nature of these thresholds has a dream quality and requires us to move into an imaginative mode, attending as if with peripheral vision or hearing unexpected whispers from another reality.

This psychology of borders has long been a strong interest of mine. Much of my formative experience was in the late 1960s and early 1970s, when social turbulence dissolved old forms and a partial cultural renaissance emerged. Recognising creative potential of margins and learning to navigate those in-between realities ('Dangerous Margins', Robertson, 2011) led to an early workshop entitled 'The Borderlands and the Wisdom of Uncertainty', the subject of a BBC documentary in 1990 entitled *An Inner Country*. Central to the workshop was the preparation for a ritual transition in which participants relinquished old identifications, which they recognised no longer served them and opened to an emergent sense of becoming. Many found the beginning of new stories that were yet to be lived.

Jerome Bernstein, a Jungian analyst from New York strongly influenced by his lengthy experience with the Navajo nation, explored a similar threshold reality in his book *Living in the Borderland* (2005). He describes the emergence of the 'Borderland,' as a spectrum of reality beyond the rational, yet palpable to an increasing number of individuals. He notes the dangers in pathologising human sensitivities, which may be diagnosed as borderline personality disorders, whereas they more accurately reflect the distorted nature of a rational ego divorced from non-human nature. The borderland is a relatively open space between human and other-than-human which allows a healing reconnection through portals of communication between the human animal and non-human animals.

What the psychology of borders illuminates is:

- the precarious nature of our alienated separate ego and the social defences it has created trap it in maintaining a need to dominate and control
- the process of relinquishing these defences is fraught with terrors of abandonment and annihilation
- a ritual container is needed to hold the unbearable feelings and allow a new dream or a new story to be birthed.

Freya Mathews, an Australian environmental philosopher, explores the making and breaking of a primary intersubjective bond with the environment and human world in *Reinhabiting Reality: Towards a Recovery of Culture* (2005). Like Bernstein, she draws on indigenous knowledge and contrasts it with western philosophy and science. To illustrate some of the wounds, reparations and wisdom explored in her comprehensive book, she tells the story of 'The White Heron' (2005, p. 115) abbreviated as follows.

> Sylvia has come from a troubled and impoverished life in the city to live in the woods with her grandmother. The country setting allows this shy, withdrawn and probably depressed young girl to begin to find herself. She loses her apathy and begins to thrive. Through a close connection with the wild birds and animals in the surrounding woods, she feels as if she had never been alive before coming to the woods.
>
> One evening she is approached by a hunter and ornithological collector, who is in the area looking for birds for his collection. This young man is particularly searching for a rare white heron, and he is sure it makes its nest in the vicinity. He accompanies Sylvia on her way and asks for accommodation at her grandmother's house. He stays and shows himself to be polite and gallant. When he acknowledges his mission of searching for the white heron, Sylvia is tongue tied because she has indeed seen this heron. Suspecting something in her silence, the hunter offers a huge sum of money to anyone who can lead him to the white heron. The next day Sylvia accompanies the hunter into the forest as he searches for the bird's nest, but he does not find it. She is increasingly enchanted by this handsome young man and longs to aid him in his quest.
>
> Early the following morning, Sylvia plans to climb the tallest tree in the forest so she can see the entire countryside and find the heron. When Sylvia makes the dangerous climb, she arrives at her dilemma at the tree's top. She does indeed spot the heron, which flies straight towards her tree and alights on a branch nearby. Knowing the heron's hiding place, she has the object of the young man's desire, but will she speak it and betray her connection to the heron? She desperately wants to win his love and approval but finds herself mute and unable to tell him. He eventually leaves disappointed and empty handed.

Freya Mathews characterises Sylvia's dilemma as this: having made a primary bond with her natural surroundings, she has little social identity and has never had to adapt her sense of self to the requirements of a human other. She has not had to define herself as an object to others in the process of socialisation. What is being asked of her in this encounter with the hunter? She is being asked to betray a primary trust with the other-than-human to gain the approval of the hunter and further her social status. The dilemma highlights the split between our socially constructed power-over culture and the deep sense of indigenous belonging.

It would be easy to read this story as this a failure of socialisation; of a refusal to emerge from an instinctual protective womb. From a diagnostic view, Sylvia might be regarded as on the autistic spectrum, but from the perspective of this chapter, her dilemma is reflective of our cultural crisis. We are in the liminal in-between zone; tempted by the promises of wealth, power and status if we align with the predator (a cypher for the exploitation of the industrial complex) or holding to an ecological sense of allegiance to our planet. Mathews writes graphically (p. 129):

> Moderns know what they want, and they move in for the kill to get it. They elbow their way forward, determined to reach their destination. They jostle for the centre stage, competing for the limelight of approval. The native meanwhile irresolutely eddies here and there, tarries at the edges, lingers in the background.

The normative cultural ego makes Sylvia the aberration, the reclusive autistic without enough socialisation. But if we reverse figure and ground, it is equally easy to see our normative view as the aberration and Sylvia's borderland sensitivities as a portal to a different cultural reality. It cannot be a return to the garden because the garden, both literal and symbolic has gone, but it could draw on indigenous earth wisdom to engage a new dream, a re-enchantment with the natural world that fosters an indwelling and sense of presence. Mathews writes (p. 129),

> The native self, by contrast, dwells within the parameters of the given. She looks for no testimonials to her existence – no obituaries or biographies – nor does she seek to recast earth in her own image. She stands in the calx of the inexhaustibly deep and poetic reality that opens its petals around her. She breasts these layered energies, gently launching herself on their swells and swirl, attuning herself to the inner rhythm, the pulse of their unfurling, letting the pulse lift her and carry her forward. She adjusts her own inner curvature to that of the world, and in gratitude she lets the waves plot her course.

So, if we are sometimes rendered mute in response to challenges on climate change, life after capitalism or post-collapse societies, it may be an appropriate pause; a refusal to be drawn into a language not quite ready to be formed; an inarticulate speech of the heart, as Van Morrison (1983) named it in an album title. This pausing is a meditative counterpoint to the cultural mania, a deceleration in contrast the technological acceleration towards potential extinction. Pausing in a liminal space, we need to exercise negative capability and refuse to fill in the space with what is already known, to stay with the uncertainty and allow the unborn to emerge in its own time. If we pause, allowing the waves to plot our course, it is likely the white heron will alight near us. It is waiting for us while we are busy looking for her!

Although climate change, financial inequality, dependence on fossil fuels, migration, the loss of species and biodiversity are all pressing problems requiring political agency to correct, they are symptoms of our greater cultural malaise – a deep sense as a species, humans have made a wrong turning and lost our way. This chapter presents the need for cultural transformation involving relinquishing our privileged superior status as a species, which has had such destructive consequences for the other-than-human and our shared planet. The challenge of this relinquishment is a scary descent through a borderland reality along the lines Jung called for almost 90 years ago. It will mean a revisioning of what it means to be human and our place on earth. Having lost our soul as human individuals, we may find it afresh through being entangled in a participatory ensouled world which welcomes us back from our estranged dissociation.

References

Abram, D. (2010). *In the depths of a breathing planet: Gaia and the transformation of experience*. Available at: https://wildethics.org/essay/in-the-depths-of-a-breathing-planet/.

Albrecht, G., Sartore, G.M., Connor, L., et al. (2007). Solastalgia: The distress caused by environmental change. *Australasian Psychiatry*, 15(s1), pp. S95–S98.

Anderson, J. and Robertson, C., eds. (2016). Climate change and radical hope. In: *The psychotherapist*, Special Issue, 63. London: UKCP.

Baldwin, J. (1963). *The fire next time*. New York: Vintage Books.

Bendell, J. (2018). *Deep adaptation: A map for navigating climate change tragedy*, IFLAS Occasional Paper 2. Available at: www.lifeworth.com/deepadaptation.pdf.

Bernstein, J. (2005). *Living in the borderland: The evolution of consciousness and the challenge of healing trauma*. London: Routledge.

Berry, T. (1988). *The dream of the earth*. San Francisco, CA: Sierra Club Books.

Bion, W, (1962). *Learning from experience*. London: Maresfield Library.

Cameron, J. (2009). *Avatar*. Directed by James Cameron, 20th Century Fox.

Camus, A. (1960). *The plague*, trans. by S. Gilbert. London: Penguin.

Costing the Earth. (May 2019). Available at: www.bbc.co.uk/programmes/m00050qr.

Diamond, J. (2006). *Collapse: How societies choose to fail or survive*. London: Penguin.

Farley, P. and Roberts, M. (2011). *Edgelands: Journeys into Englands' true wilderness*. London: Random House.

Franzen, J. (Sept 8, 2019). Available at: www.newyorker.com/culture/cultural-comment/what-if-we-stopped-pretending.

Hill, M. (2005). *Dreaming the end of the world: Apocalypse as a rite of passage*. Dallas: Spring Publications.

Jung, C.G. (1953–77). *Except where indicated, references are by volume and paragraph number to the collected works of C. G. Jung*. 20 vol., ed. by H. Read, M. Fordham, and G. Adler, trans. by R.F.C. Hull. London: Routledge and Princeton: Princeton University Press.

———. (2001). *Modern man in search of a soul*. London: Routledge.

———. (2009). *The red book: Liber Novus*, ed. by S. Shamdasani. London: Norton.

Klein, N. (2008). *The shock doctrine: The rise of disaster capitalism*. London: Penguin.

Lawrence, G. (1998). *Social dreaming at work*. London: Routledge.

Lear, J. (2006). *Radical hope: Ethics in the face of cultural devastation*. Cambridge, MA: Harvard University Press.

Lorenz, E.N. (1993). *The essence of chaos*. Seattle, WA: University of Washington Press.

Lovelock, J. (2006). *The revenge of Gaia*. London: Allen Lane.

Mathews, F. (2005). *Reinhabiting reality: Towards a recovery of culture*. Albany, NY: University of New York press.

Manley, J., Gosling, J. and Patman, D. (2015). *Full of dreams: Social dreaming as liminal psychic space*. Available at: www.researchgate.net/publication/289470156_Social_Dreaming_as_liminal_psychic_space.

Menzies-Lyth, I. (1970). *The functioning of social systems against anxiety: A report on a study of the nursing service of a general hospital*. London: Tavistock Institute of Human Relations.

Morton, T. (2018). *Being ecological*. London: Penguin.

Robertson, C. (2011). Dangerous margins: Restoring the stem cells of the psyche. In: M.J. Rust and N. Totton, eds., *Vital signs*. London: Karnac.

———. (2018). *Transformation in troubled times*. Forres: Transpersonal Press.

Romanyshyn, R. (2019). *Victor Frankenstein, The monster and the shadows of technology: The Frankenstein prophecies*. London: Routledge.

Turner, V. (1967). Betwixt-and-between: The liminal period. In: *Rites de passage; The forest of symbols: Aspects of Ndembu ritual*. Cornell, NY: Cornell University Press.

Van Morrison, G. (1983). *The inarticulate speech of the heart*. London: Warner Bros.

Wallace-Wells, D. (2019). *The uninhabitable earth*. London: Allen Lane.

Winnicott, D.W. ([1971] 1991). *Playing and reality*. London: Routledge.

Chapter 6

The Golem

An image for our time

Bernard Sartorius

What I am going to write about is on the level of the so-called 'cultural complexes', on the level of a part of the unconscious which has to do with the unconscious of a given culture and which probably does not reach as deeply as the psychic 'layers' which go down to the psychoid archetypes. The Jung/Pauli dialogue (see Chapter 11 of this volume) concerns matters which might relate to that deepest level, while this seems to belong to a more accessible part of the collective unconscious. But, as we shall see, it nevertheless expresses itself in a most dramatic way because it is pointing towards the end of our technological-economical civilisation, or at least towards a dramatic crisis in its progress.

Before coming to the main topic of the Golem figure, I want to confront you with some hard facts. I believe one of the important tasks of our thinking today, especially in psychology, must be to look at what is happening to the world around us and at the way we live in this world. We cannot confine our attention to our own individual introversion. The Swiss government has recently published statistics which set the sustainability of our Swiss way of life against the capacity of the planet to absorb and support it. There are two related issues: the exploitation and using up of natural resources, and the capacity of the planet to dispose of or absorb the waste which our civilisation produces. Between 1961 and 2006, in the short period of less than 50 years, there has been in Switzerland a rapidly growing discrepancy between our demands and the capacity of the planet to meet them in terms of the provision of resources and disposal of waste.

In terms of ecological consciousness and 'green attitudes', Switzerland is one of the most enlightened countries on the planet, distinguishing between different types of waste and selecting the most biologically and organically friendly methods of disposing of its various waste products. Nor is the country over-consuming its resources. So we have in the case of Switzerland, to a certain extent, an example of a good ecological management culture. Despite this, we have this huge problem of sustainability, and this can be observed all over the planet. The issues become even more frightening and dramatically evident if we consider the explosion of productivity and development in countries like China and India.

The important point is that the discrepancy between human demands and planetary resources is growing and growing in an almost exponential way. Presently, the planet's bio-capacity (its capacity for renewal of resources and absorption of wastes) is already out of balance with our demands. Meeting what is currently needed and absorbing the waste of what humanity uses, currently requires 125% of the planet's resources. I repeat – we are already at the point of needing 125% of the planet's resources to maintain our current collective lifestyle! If this trend persists, then by 2050 we would need two planets to sustain us – two planets! So, of course there is a solution: we can send half of humanity to Mars or the Moon. Maybe that idea accounts for some science fiction fantasies but, of course, it provides no solution.

We have here a 'development' in our material progress which isn't an evolution. If we look diatonically, that is, extensively, at historical time, without even considering the far greater geological time behind us, we must regard what has happened during the last 100 years as an explosion. And it could very well prevent further evolution. It is very important that we feel the difference of quality between natural progress – evolution – and what characterises an explosion. The progress of a cancer is indeed organic, but it is an explosion. It is no longer a natural progress, an evolution, for an evolution relates to the whole organism. In cancer, we witness an explosion of cells which do not anymore fit into the organism: it is an intrusion of something very much alive into the living whole, thriving itself but eventually killing its host. At the level of the development of civilisation, we seem to have a similar process: our technological – economical civilisation is very much alive and thrives on natural resources but nevertheless, in the long run, it is deadly for life itself.

My colleague Wolfgang Giegerich spoke some time ago about the atomic bomb. It was, and still is of course, a real danger. He further showed the atom bomb is symbolically important precisely because it constellates the fear of a major explosion which would threaten all of life. This conscious fear of the atomic bomb could well mirror our mostly unconscious anxiety in front of the 'cancerous' explosive growth of our civilisation.

My hypothesis is the roots of this explosive development and resulting ecological crisis lie in Western cultural complexes. The basic roots of the evolution of civilisation are, of course, human – mankind has always wanted, among other things, to develop technologies in order to ease life, but this drive gained a potent, explosive Western component: a combination of the ancient Greeks' discovery of the importance of rationality with Judeo-Christian anthropocentricity – man seen as the centre of everything, particularly because of his superior rationality. It was in the Renaissance when this merging received its most powerful momentum, leading to the Enlightenment and to 'modern times'.

It is not necessary for me to describe the roots of rationality developed in ancient Greece by Plato, Aristotle and others and the fundamental discovery of individual thinking. About the Judeo-Christian component of this Western complex, I want to say a bit more. If we consider Judaism, we see a key feature: God became involved in history. By the way, when I speak now of 'God', I am speaking of the image of God. I cannot speak about God himself, for, as Jung pointed out, God himself or itself is not a matter of psychology. I can speak only of the image of God. The image of God in Judaism evokes a god who was interested in the history of mankind and particularly in one group of people, the Hebrews. Through his emotional investment in the Hebrews, he intervened in history, educating them and bringing them to the Promised Land. And further – another important feature – God was seen as being radically distinct from nature.

As you may know, in the Old Testament, there was a permanent fight between the Hebrews and tribes which worshipped nature gods. These pagans were worshipping the divine in nature whereas the God of the Hebrews, Yahweh, insisted on an absolute separation between himself and nature. Here I want to refer briefly to Islam, another monotheistic religion. In Genesis, the relationship between man and creation is 'man may rule over creation' (Genesis 1:26). The Koran, on the other hand, says that 'man is the "Caliph" of God' (Sura 35:39), the 'representative' of God in the creation. This difference in wording is more than a nuance. It indicates that Islam is in a certain way psychically close to nature religions: man has to take care of the creation as a representative of God and is not given the right just to help himself to its riches nor to rule over creation to meet more than his own basic needs for life.

In Judaism, God is more interested in humans than in the rest of creation. In Christianity, this image of a God being primarily interested in humans becomes even stronger, for in Christianity, God becomes human in Jesus Christ; the divine himself becomes human. God does not become stone or flower or animal or sky or whatever. He becomes human. We have here a fundamental anthropocentricity, positioning humans mythically in the centre of the universe. As long as God was alive in the Western collective psyche, this anthropocentricity could to a certain extent be counterbalanced by faith in God, but with the Renaissance gradually replacing religion with reason and then with the Enlightenment ultimately leading to human reason becoming God – in Paris after the Revolution they erected a statue dedicated to Reason – this compensation for a fully anthropocentric relationship to reality disappeared and what remained was the 'pure' anthropocentric perception that humans, with our desires, are the most important reality in this world and that- this is derived from the ancient Greeks – it is through our rationality that we assert this position as rulers over nature, over the reality of life.

This myth – that we are the most important reality in the universe – leads to modern human desires being invested with absolute importance, because it is on their satisfaction the quality of subjective experience and the sense of the value of one's existence depends. These desires range upwards from material goods and pleasures to the highest aspects of spiritual well-being, to peace of mind, peace of soul, whatever. But the outcome is always more questions, such as, 'How am I? How do I feel? What is the state of my soul? How satisfied am I on the material level?' Ours is an anthropocentric culture; psychologically speaking, it is an egocentric civilisation.

I am not now discussing 'pockets' where non-egocentric spiritual values prevail, still to be found, for example, in some churches or monasteries or other

communities. I speak of the general trend of our technological-economical culture. Of course, we have had warning signs it could be heading into a dead end.
The analytical psychology movement is, itself, a warning sign as it points to an
'unconscious' not reducible to human desires and Jung's concept of the 'Self' is
not reducible to applying to humans alone. Moreover, if we look at the evolution
of art in the twentieth century, we can see signs of a psychic movement beyond
anthropocentricity in the works of artists who are expressing abstraction and
eventually dissolution. In those works, humans are not at the centre anymore, not
in charge of things, and one sees there is something – a compensation for this egocentricity – emerging, a compensation beginning with dissolution and emptiness.

Then there was the warning sign embodied in the regressive political movements of the twentieth century, when non-egocentric values apparently started to
reappear. I am referring to National Socialism, fascism and movements preaching
values such as 'blood and soil' or 'blood and race'. Through them, a resurgence
of the divinisation of nature took place but – as always when a repressed content suddenly irrupts into consciousness – in a crazy, pathological and destructive
way. In communism, values such as 'the others', 'the community' and 'solidarity' were also used to outweigh and supersede egocentricity but, like 'back to the
roots' political compensations, these extremist social movements had to collapse
because they were perverted, being nostalgic attempts to revert to previous times
of lesser egocentricity. They were signs of a pathological regression resulting
from modern humans seeing themselves in the centre of everything.

Now we are in a vicious circle: on the one hand, there is a tremendous need for
personal satisfaction which is to a large extent materialistic because this 'mater'-
ialism is linked symbolically to the lost contact with 'mother' nature, a loss felt
deeply in the unconscious. At the same time, there is a widespread and basic
uneasiness with what is going on because everybody, in a confused way, feels that
this material 'progress' is going nowhere, this is leading into catastrophe. And
that's where the vicious circle closes itself: in order to alleviate this uneasiness,
most people compensate with even more consumption of material goods . . . and
so it goes on.

The only people who do not yet feel this unease are population groups on the
fringe between the 'developed' world and cultures still in close relationship with
nature. People coming from rural areas and from so-called 'developing countries'
still do not feel the tension because they have been so recently exposed to possibilities of satisfying their desires and making life 'easier', and these possibilities
seem to their eyes to be unlimited. Most probably, they will nourish this way of
functioning for the coming decades, producing new generations which will gladly
contribute to over-use of the resources of the planet. Thanks to large populations
in emerging countries, there will be enough energy and people wanting this dangerous train to stay on its tracks for a long time to come. Yet the dead end, the
wall, is rapidly being approached.

Today, we witness other compensatory developments which are not ideological, for instance, tourism: mass tourism lures huge populations into so-called

'unspoiled' areas, rapidly destroying them. I fear those articles you find in newspapers about a not-yet-discovered island, or a not-yet-discovered beach. Two or three years later, tour operators are planting their buildings there. Mass tourism is a good illustration of the cancerous nature of our civilisation as it spreads like metastases. With few exceptions, it is not organic and it eventually destroys the 'body' – landscapes, local cultures – it is feeding upon.

All these observations cannot but leave, eventually, a feeling of meaninglessness. This meaninglessness is, to a large extent, the spiritual condition of modern humans, a condition in which even 'deep' searching for meaning, such as through Jungian ideology, can become, as Giegerich shows, a way of not seeing reality, an escape into a too much individualised, imaginary and personal world. I believe many contemporary dreams – but this could be the topic of another paper – speak to our failure to be consciously connected with what is happening in present-day civilisation.

Now let's go to the Golem, because you will see, especially after the description I have tried to give of the dead end of our present civilisation, this legendary figure is incredibly prophetic. I shall take one of the many Golem stories, but if you are interested in this motif, there is a monograph on the Golem by Gershom Scholem in the *Eranos Year Book* 22, 1953. I have taken my material from this essay which is a complete anthology of the Golem motif. I chose a story from this paper, one of the more recent ones Scholem found, a story collected by Jacob Grimm, one of the Brothers Grimm, in the early nineteenth century. This story has most of the features of the Golem including the last developments of the motif. I quote:

> The Polish Jews, after appropriate prayers, used to make out of clay or mud a human figure and when they pronounced the name of God over it, it became alive. This figure could not speak but understood the orders it was given. They called it "Golem" and used him to do all kinds of work inside the house, but he should never step outside. On his forehead the word "emeth", which means truth, was inscribed. It is in the nature of the Golem to grow a bit taller every day, and thus to become stronger than the other inhabitants in the house. In order to prevent that happening, they used to erase the first letter of "emeth" which thus becomes "meth", meaning dead or death. So, the Golem dies: he collapses and returns to mere clay. However, one story tells of a Jew in Krakow who had carelessly let his Golem grow to well over his own height so that he could not any more reach the Golem's forehead in order to deactivate him. So, he ordered the Golem to help him take off his boots, knowing that he would be able to reach the Golem's forehead when he bent over. That is what happened and as the man erased the first letter of emeth, the Golem indeed collapsed. However, his whole weight of clay fell upon his owner and crushed the man.

> (See also Scholem, 1969, pp. 285–6).

The word 'Golem' in Hebrew designates a reality that is unshaped and it can signify an embryo, something unfinished, something without spirit. In this sense, Adam is described as 'Golem' in some texts of the Midrash, in order to distinguish human creativity from God's creativity. Whatever humans do, we do it like a Golem in a crude, non-'spiritual' way, in comparison to God's creativity. In other words, humans have a 'Golem' nature because we are not as spiritual as God. 'I want', says God in a text of the Midrash, 'to leave man in the state of a Golem until I've finished creating the world'. So, the Golem aspect of humans symbolise our status as unfinished and of imperfect nature in comparison with the perfect nature of the true creator, Yahweh.

The hubris of imagining oneself as being a real creator is thus archetypally not possible. In this Judaic tradition, humanity has Golem nature in comparison to God; in our nature, we cannot be true creators in comparison to God's creativity, though on the level of our conscious ego, we think that we can be truly creative. As I said earlier, in general, Judeo-Christian Western civilisation is anthropocentric, but we see already in this attribution of a Golem quality to Adam a compensatory intuition: human creativity cannot compete with God's creativity.

A second point about the nature of the Golem in this story is he is made of clay, of mud. This is to say he has a material quality; he is of matter. Being shaped out of clay, there is something earthy in him, so he might represent the human need to experience the concrete, the material, something one can touch. It is what I can hold which gives me security because I can hold it and that includes, of course, all possible material goods and technologies. Matter as such is, symbolically speaking, in relationship with the earth, connected to the mother archetype and so to be made of clay can mean that the Golem has a motherly quality. On the other hand, the Golem is active – does housework, washes the dishes, sweeps the floor, whatever; that's probably what he was used for. He is active and at the same time he is made of earthy clay, thus combining masculine and feminine qualities. The masculine component is his activity and the feminine component is the earth of which he is made.

One could imagine here a conjunction between masculine and feminine qualities. But let's have a closer look at this coniunctio: could we not in a similar way, say that electronic language, being made of 1 and 0, being constructed of masculine components and feminine components, expresses a coiunctio? But if we imagine here, and in the Golem, a coiunctio as Jung wrote about it, an integration of basic living opposites, would it not leave us with a strange feeling? The Golem is for sure a combination of something archetypally feminine – the clay – and something archetypally masculine – his activity resulting from the words *emeth* (truth) or *Jahve* (God) put on his forehead – but is it a living conjunction as symbolised for instance in alchemical texts by the marriage of the king and the queen? That's the question.

There can also be 'monstrous' coniunctios. Can one say there is soul in the functioning of the Golem? He has indeed a masculine and a feminine component which are activated but not *alive*; it is difficult to imagine that he has soul. This could be one of the reasons why the Golem has no limit in his growth. There is no anima, name it as you wish, no soul. There is no relatedness. Thus, he embodies a monstrous coinunctio. The Golem has only growth and this growth follows its own momentum because – there we have again the cancer motif – he is not linked to life as such; he is not *erotically* connected to his surroundings; in his very being he is 'autistic'; when ordered to do something, he executes it literally, mechanically.

Consider how this Golem comes 'alive', becomes activated. He becomes activated through the name of God put on his forehead, or through the word 'truth'. In fact, this 'truth', according to the Talmud, is the 'seal of God'. So, this 'truth' is very strongly connected, if not identical with God. In other words, by animating the Golem by inscribing 'God' or 'Truth' on his forehead, his maker projects something divine onto this clay figure, that at first glance makes him seem alive but in fact makes him merely functional. He is not really alive because that would mean he has soul, but he can indeed function and has the capacity to obey orders. It is by putting onto him this projection of something absolute – God, the Truth of God – by projecting the *absolute* on him that the Golem can function. This seems to be consistent with what I said about the anthropocentricity of Western civilisation: when I produce a Golem to fulfil my desires, I project something 'divine' onto this instrument because I give to my desires an absolute, *divine* quality. My egocentric desires inscribe the name of God on the Golem's forehead because – of course without being so conscious of it – I worship as divine the fulfilment of my desires, hence the tools helping to fulfil them. This huge projection makes the Golem functional.

It is important to note at this point that the earliest Golem stories, dating from the Middle Ages, gave to this figure another function. The first Golem stories describe mystics, *hasidim*, making little clay figures who, in a mystical contact with God, were inscribing the name of God or *Emeth* on his forehead. Then this small Golem would become alive and start to move, but this was done, say the texts, *only* in order to celebrate the power of God's creativity working through the mystic. In other words, humans, beings in the image of God and very close to God in a mystical state, could miraculously make a little Golem so God could demonstrate His creative potential. Once they had made this demonstration and the little Golem was walking around on the table, the hassidim stopped him.

In those early texts, there was not the slightest intention to use this Golem for humans' purposes. It was only an illustration of God's power at work in a miraculous way through a human being able to make a little Golem if they have an intimate and intense connection with God. There is no mention in those texts of any tendency in the Golem to automatic growth. But then, later texts dating from the Renaissance – and this corroborates what I said about anthropocentricity intensifying during this period – describe the Golem as being made only in order to become an instrument, and that is when those stories appear of the Golem helping in the house involved with the risk of his becoming bigger and bigger and the consequent need to stop this growth at some point.

Let's return now to our particular Golem story. In order to stop the Golem from his continuous growth, it is said his maker had to erase the first letter of the word *emeth* 'truth', the Truth of God – thus producing the word *meth*, which means dead. This being dead can be interpreted as referring not only to the Golem, which then collapses back into mud, but also to divine truth becoming dead. In other words, if the word for God's 'truth' becomes the word 'dead', this can be read symbolically as the death of God's truth, hence of God himself. Here we might spot an anticipation of the whole mythology of the death of God which Nietzsche was to proclaim a few centuries later. Without going so far for the time being, we have in the eradication of God's name in the Golem stories of the Renaissance at least a premonition of the end of the image of God being instrumental, of a God who prevents evil, heals people, an instrumental God who does useful things, such as preventing suffering. This representation of a God who is doing good things according to our human wishes has indeed collapsed in our civilisation, even as we – or Jung in his *Answer to Job!* – resist acknowledging it. As Jungians, we tend to divinise the unconscious and here we have an indication *any* instrumentalisation of the unconscious – 'listen to your dreams and everything will be all right' – might come to an end, or is already, to some extent, at its end.

I have worked for 30 years as an analyst and I must say I often have had the impression the unconscious does not so much want to heal people, in the sense of achieving psychotherapeutic goals, as to open their – and our! – eyes. When we analyse the particular episode in our Golem story of this man forgetting to deactivate the Golem in time, we come very close to the dangerous collective situation we have evoked. The story says the man was so full of enthusiasm for the good things the Golem did in his house, cleaning it, being so helpful to his wife, that he forgot to deactivate him. It was his fascination with this most useful work done by the Golem which caused him to forget the moment to stop him. Remember how the story ended. When the Golem had become too big to be touched on his forehead, his maker made him bow down, made him take off his boots. So, then the man could reach the Golems forehead and 'deactivate' him. He succeeds, but the weight of the collapsing Golem crushes and kills him. We are here in an utterly dramatic situation.

The story shows a dilemma that has no solution: either his maker lets the Golem grow further and destroy the house, not because he is evil but just because of his growing size and his movements, or his maker succeeds in stopping him but then the Golem collapses and kills his maker at the same time. When we discuss big ecological problems, we are today in this sort of dilemma. Do we want our economy – the collective 'cancer' I spoke about – to collapse? Of course not, but in fact, as everything is now connected, if we really want to save the planet, in a global sense, and stop the technological-economical cancer, we would have to lay off I don't know how many millions of people from work. It is a basic, unsolvable dilemma. That is why for me, this particular Golem story is so important – either the Golem destroys the house or the man is crushed by the deactivated Golem – because it does not allow a too easy, shallow optimism.

Before thinking about solutions to our current ecological dilemma, we have first to be fully aware of the situation. Any kind of thought about concrete solutions right now is, to my mind, premature and this is, of course, very hard to accept. It is not an easy psychic situation and that is why most people prefer not to think about it. The former United States Vice President Al Gore and people like him say we can keep our lifestyle and everything which goes with it and that, at the same time, we can save the planet. This does not seem realistic. As I said earlier, the most recent statistics and facts are against such optimism.

Before making some tentative conclusions, I want to discuss another Golem story told by Scholem (1953) which, if it does not show a concrete way out of the dilemma described here, at least evokes the fantasy there could be a symbol of a solution.

> There was once a rabbi in Prague, a famous rabbi, Rabbi Loew, who had made a Golem to be used in his service during the week and every Sabbath he very wisely "deactivated" it. However, one Friday evening he forgot to do this and as the first prayer in the synagogue was being said, the worshippers heard a terrible noise outside: the Golem had become too big, had broken loose and was destroying the neighbourhood. Rabbi Loew rushed outside and managed by jumping at him to remove the name of God from his forehead and deactivate him. This was only possible because it was still dusk and the Sabbath had not really begun; had the Sabbath begun it would not have been acceptable for him to do this. Then he went back to the synagogue and ordered the first prayer be said again; he wanted the service to be totally coherent. And in this particular synagogue in Prague this prayer was still – until the Second World War – said twice because of this story.

There are two symbolically interesting elements in this story. The first is that Rabbi Loew made a *spontaneous* jump. He did not wait a second and the jump was unexpected and unconventional to the extent of interrupting the sacred act of prayer. The only thing we can say about the present dilemma of our civilisation is that if there is no spontaneous 'jump' disrupting the 'sacred' rituals of our time (economic and social habits, this is!), then the dilemma the Golem story illustrates might well take its course, either towards destruction of the planet or of part of humanity. I cannot say what this jump would mean in practical terms, but the action of Rabbi Loew might be taken symbolically to represent full psychic mobilisation and quick action, not waiting for solutions but engaging in an immediate breakthrough to stop the Golem.

The second aspect of this story is that, by having to jump at the Golem, Rabbi Loew left the synagogue: he went out of the sacred space. At that crucial moment, he disregarded the synagogue which could be taken to symbolise what was most 'interior' for him. Action was necessary and *only* action and it was extraverted action, not introverted experience of the sacred space. So, our contemporary Golem dilemma could well constellate a need to act, to *do* what subjectively seems right, without taking time for long introverted reflection and meditation.

Now just a few conclusions . . . well, conclusions are impossible, but I should like to identify some elements in what we have considered and how they might bear on our work as psychologists, Jungian enthusiasts and so on. First, the unconscious of more people than we think in our modern world does know what is going on. In the unconscious there are dreams which show what is going on. I have quite a few patients whose dreams reveal the unconscious is feeling what is going on and is not happy about it. This is an objective fact. An analysand brought a dream, which he has given permission to share, as an example:

> There is an important building made of iron and glass, like the Pompidou Centre in Paris with all its pipes visible, and a huge turtle is sleeping in the basement of this building. Suddenly a pipe is leaking and acid starts to drop on the shell of this turtle. At the beginning the animal doesn't feel anything because the carapace is strong enough but then suddenly a hole is made; the acid has worked through the shell and created a hole. The turtle feels the acid painfully and starts shaking. This shaking produces an earthquake and then the building collapses. After a while, the turtle emerges from the rubble, its shell intact and leaves the scene by moving towards the right side of the picture, which is usually interpreted as indicating the future.

This dream presents a deadlock, destruction and collapse, but life goes on. And the turtle, which can stand for so many aspects, can be the planet itself, for in old mythology – for instance in China – the turtle sustains the weight of the universe and is as an image of life as such. This turtle will react, or is maybe already reacting, and will shake and destroy the building, but she will survive. The building is a fabrication, nothing permanent, merely a transitory construction, an image of civilisation.

Second conclusion: we know the collective unconscious can affect us as individuals and often when something is unresolved in our culture, the personal problems which people have are fed, so to speak, from below by energy coming from this collective uneasiness. Problems of personal relationships become bigger; the ways personal difficulties at work are experienced become inflated and so on. There are, for sure, real personal problems, but there is also an energy feeding them from below and today particularly such energy comes from the Golem dilemma I have been describing. In my work I have come to realise that it is important to connect people to such cultural problems so they can relate their personal pathology to these collective issues.

Third conclusion: this Renaissance story of the Golem is, as I have tried to show, a 'tremendous' (in the sense of *mysterium tremendum*, a mystery producing fear) illustration of our present collective cultural situation and it thus challenges major political and economic priorities of our time. It shows an archetypal image or scenario appearing in legends and myths can have a huge *collective* bearing: the impact of the archetypal goes far beyond the life of the individual.

We see today in the media intense apocalyptic fantasies such as in the many movies about the end of the world produced by Hollywood and other studios. They depict violence threatening and destroying humanity. They appear also on the individual level, where this collective sense of an ending flows into the death fear and fantasies of the individual. All this might indicate we are going towards an unknown future, a *really* unknown future. We cannot exclude the possibility this basic psychic uncertainty about the future is pointing towards God as an aspect of what we call 'God' might be constellated through the fact that the future is unknown. Theologians understand a basic aspect of God is that He is unknown. So the experience of the unknown, the existential experience of the unknown, the fact I cannot project a clear vision of what's going to happen tomorrow or further in the future, can be the foretaste, let us say the portal, to a new religious experience. It may not be a spectacular religious experience, but the experience of the unknown can be a door towards it.

If we recall the end of our first Golem story, with the unsolvable dilemma facing the maker of the Golem of either having his house destroyed or himself crushed, destruction and death could be such a door. There is today a widespread interest in religious phenomena among many people who are not positioned to relate to the traditional ways in which religion is transmitted. What does seem evident is this interest in religion points towards a religiosity which is not anymore anthropocentric – religion which wants to maintain the human being with 'salvation' and well-being as the most important reality in the universe is *passé*.

Al-hamdul'Illah.

References

Scholem, G. (1953). *Die Vorstellung vom Golem in ihren tellurischen und magischen Beziehungen.* Eranos-Jahrbuch Band 22, Rhein-Verlag, Zurich 1954, pp. 235–289.

———. (1969). *On the kabbalah and its symbolism*, trans. by Ralph Manheim. New York: Schocken.

Chapter 7

Imagining earth

Jules Cashford

Both Jung and Einstein, among many others, agree that no problem can be solved by the same kind of thinking which created it. Einstein warned that 'With the splitting of the atom everything has changed save our mode of thinking, and thus we head towards unparalleled catastrophes' (Calaprice, 2000, p. 184). This helps explain why, in the face of a global crisis of unthinkable devastation, the overwhelming response of many of us, and certainly almost all national governments, has been to do absolutely nothing – almost as though it was impossible to imagine what to do. Jung added his own terrifying image: 'The world hangs by a thin thread, and that thread is the psyche of man' (Jung, 1959).

The urgent question then becomes: what was, and still is, the thinking that has brought about this crisis, and can we assume that we will even be able to recognise it? We might suspect a kind of thinking so powerful as to destroy life forms all over the planet issues from a distinctly specific and long-established view of the world, one we have been thinking in for so long that it now 'feels natural.' Yet we also know no way of thinking exists in a vacuum. Rather, it comes from an almost inconceivable plethora of other thoughts and feelings and, more crucially, the values which underpin them, the way of life which comes out of them, and, ultimately, who we are as human beings, living together on our one and only Earth.

If this is so, then the members of our culture will not simply be able to 'decide' intellectually that our former world view, once so convincing, is now 'wrong,' and resolutely pledge to give it up. Would we not first have to understand where it came from and why we have failed to question it for so long? Let us see if another culture with a radically different relationship to their Earth may offer us a perspective on our own.

In the Imagination of Ancient Greece, Earth, *Gaia*, 'Mother of All, the first to arise from chaos,' was always honoured as a living presence whose laws were written into the lives of all creation to whom she had given birth (Cashford, trans. 2003, p. 140). It followed that Gaia's law was related to the moral law of human beings, or, as we might now say, Nature and human nature at the deepest level were not separately configured. In other words, the order of Nature was for the Greeks a dynamic moral order, both an agent and a reflection of human life. The first king, Erechthonius was 'crowned' by Gaia, bringing him up as a baby out of her dark depths into the light. Just as, in ancient Egypt, two thousand years before, the Pharaoh received his right to rule by being seated on the 'Lap of the Goddess Isis' as his throne.

It followed that Gaia's law could be profoundly disturbed by the immoral and unlawful behaviour of human beings. Sophocles, in his *Oedipus Rex*, draws an Earth intimately related to the moral life of humanity. Oedipus, the king, is quite content in his unconsciousness until Earth suffers. Suddenly, the land of Thebes begins to die:

> A blight is on the fruitful plants of the earth,
> A blight is on the cattle in the fields,
> A blight is on our women that no children
> are born to them.
>   (Grene and Lattimore, trans. 1960, p. 112)

It is Gaia's protest which initiates the drama of Oedipus's awakening to who he is and what he has done. Oedipus sends to the Delphic Oracle of Phoebus Apollo to reveal the cause: the slaying of his father and the marrying of his mother. And the oracle, whose first law was 'Know thyself' and whose second law was 'Nothing in excess,' now defines for all time the meaning of *pollution* as a human crime against the divine order, the profaning of what is sacred:

> King Phoebus in plain words commanded us
> to drive out a pollution from our land,
> pollution grown ingrained within the land.
>   (Grene and Lattimore, trans. 1960, p. 114)

The word for pollution in Greek is *miasma*, from *miaino*, to stain, coming into Latin via *polluere*, to soil, defile, contaminate, and further back from Proto-Indo-European (PIE) *leu*, to make dirty. All these terms draw on the original archetypal distinction between clean and unclean, which carried the further meanings of sacred and profane. The Greek word *Katharsis*, from the verb *Kathairo*, to cleanse, draws on these dual meanings, as does the 'immaculate' conception of the Virgin Mary, coming from the Latin *immaculatus*, meaning literally 'unstained,' 'not spotted or defiled,' in body and spirit.

Here, the pollution is the presence of the unwitting murderer of Laius, the former king and Oedipus's father. When Oedipus discovers he is himself the pollution and leaves the city, harmony between the human and divine order is restored and Earth comes back to life. It is significant how, even without Oedipus's intention to do wrong, pollution occurs. Later, in reverse, in *Oedipus at Colonus*, the place, Colonus, where the older and wiser Oedipus is to lay his body in the Earth, will bring blessings to the people who live there. Again, Gaia and the children of Gaia are shown to be profoundly related.

So here, what happens to humans happens to Earth, and what happens to Earth happens to humans – the soul of the one is also the soul of the other. Or, put another way, the human story and the universe story are one and the same. And without an innate harmony between them, both suffer – Earth and all creatures on Earth, including human beings. With so rich and complex a tradition, who would have suspected this was the last time in the West the Earth was formally revered as sacred – nearly 2,500 years ago? Even to talk, then, of the sacredness of Earth is to press against the weight of 2,000 years of religious and cultural history, together with conscious and unconscious assumptions about the nature of reality.

Perspectives on the evolution of consciousness

The philosopher and literary critic Owen Barfield (1898–1997) has offered a unique perspective on the evolution of human consciousness which has led us to this point, and suggests a way of looking at our history which may free us to imagine a way out of our present impasse. In his *History in English Words* (1967), *Saving the Appearances* (1988), and *Poetic Diction* (2010), Barfield understands the evolution of consciousness as falling into three distinct stages of different kinds of what he calls 'Participation' with the world. To begin briefly with his definitions, which will be explored more fully later:

The first stage, existing from the beginning, he calls 'Original Participation,' which he describes as an *instinctive union* with the Earth and all her creatures, where the human individual soul and the Soul of the World are experienced as one whole. As in *Oedipus*, the human and the natural world, being indivisible, cannot be conceived in isolation from each other.

The second stage Barfield defines as a *Withdrawal* of Participation from the Earth, when the individual soul separates from the World-Soul. Earth then loses numinosity and is set in opposition to humanity, while the lost numinosity of Earth is then transferred to the inner life of human beings. At its best, this leaves humans freer to explore the complexities of their own specific natures, but at its worst allows humans to exploit the now 'soul-less' Earth for human ends alone. The inevitable danger was withdrawing our instinctive participation from Earth meant not only would we identify exclusively with ourselves, but also specifically with our new experience of an independent inward life, no longer at one with the outward Earth. So we could no longer respond to, and learn from, our so-called 'animal instincts' – our longings, passions and terrors – which fell under the opprobrium of being too susceptible to the now banished non-human world. So, while we have finally become aware of the devastating consequences for Earth, and ultimately for ourselves as 'Children of Earth,' we may not realise we have also been distorting our own consciousness which may still carry the loss of 'the source' in some deep place of the heart. This may be another reason why we feel hopeless and helpless.

The third stage Barfield calls 'Final Participation.' He defines this as a new kind of participation with Earth, not in the old, original way – which in any case is impossible, consciousness inevitably moving on – but at a new *level* through 'the Imagination.' This involves, he explains, a *dual* relation to the world, which acknowledges our present experience of Earth as separate from us, but creates a new poetic union by participating with the natural world, consciously *and* imaginatively. The aim is to bring about a new kind of relationship with our Earth, recognising our essential identity while exploring the specific human role of consciousness within Earth. And he sees this as the unique moral opportunity of our time, offering a way to transform both of the earlier stages, reconciling them in a new whole.

All these stages of consciousness are understood as necessary, on an analogy of children 'growing up.' Yet while they are presented as *historical* phases, they are also phases inherent in all human consciousness. As we learn from our dreams, nothing in the psyche is ever lost; only our understanding of it may grow and change, and while *we* may forget, the psyche does not. So although Original Participation, for instance, is placed in the distant past, and Final Participation is suggested as the future, with Withdrawal of Participation mediating in between, all three possibilities exist within us in different degrees, and one or another may take precedence at different times. May not all of us, on occasion, return to 'the infancy of the world when everyone was a poet and language itself was poetry' – Shelley's evocation of Original Participation by a different name (Shelley, 1995, p. 250)? It is only the experience of the numinous, Jung wrote, which brings about transformation in a person (Jung, 1973, p. 1, 377).

As we are now in a late stage of Withdrawal of Participation, we are at least in a position to look critically upon its limitations, and so also upon the limitations of ourselves at this point in history, even while recognising that we are only able to do this because we have in fact already withdrawn participation from the whole. This offers a unique opportunity to consider whether this present state of withdrawal should be seen as a *phase* in the evolution of consciousness, and so *not* the only and ultimate way to relate to the world. In order to do this, it seems essential to 'get behind' some of the ideas which have been taken as absolute.

Original participation

We might already understand this stage of human life through the familiar term '*participation mystique*,' coined by the anthropologist Levy-Bruhl; also from Palaeolithic cave art and the continuing way of life of Indigenous Peoples. More precisely, Barfield writes: 'The essence of *original* participation is that there stands behind the phenomena, *and on the other side of them from me*, a represented which is of the same nature as me' (Barfield, 1988, p. 42). This is not accidental or mechanical but psychic and voluntary, whether we call it the spirit world or mana or by the many names of 'God.'

The time of Original Participation was originally identified with the lunar cultures of the Goddess, where all children of the universe – animals, plants, and humans – came from her body, and, being of the same substance, were related to each other, as in the Sumerian culture of Inanna-Ishtar and in early Greek thought. The world of early people was a Thou not an It; a presence both numinous and personal, and so a Subject in the dialect of thinking, not an inanimate object of thought. All life forms belonged to the same continuum of feeling and were related through imaginative sympathy: they did not have to be apprehended by different modes of cognition. What, in contemporary terms, we would now distinguish into the objective, natural world, and contrast with the subjective, human world, were inexorably bound together, so Earth was both more awesome – loving and terrifying – and more personal – peopled with divine presences. This is imma-nence, whereby the visible appearance and the invisible source are one and the same. For the ancient imagination was concrete, as was the origin of language, embedded in, and rising out of, deeply lived experience, similar, in this respect to poetry of all ages. This union was either destroyed, or gradually diminished, at different times around the world, though surviving in Britain in an attenuated form up to the 17th century.

Withdrawal of participation

Broadly, in the second phase, the Soul of the World lost its numinosity – beginning in the late Bronze Age roughly around 2000 BC, when nomadic tribes con-quered the native agricultural communities and withdrew the immanent divin-ity from Earth, placing it either in the patterns of the heavens or in the invisible world, transcendent to all creation. Mythologically, this was the time when Gods of the Heavens – sky, storm, or sun – replaced Goddesses of the Earth – Enlil and then Marduk in Mesopotamia, Yahweh-Elohim in Canaan. Typically, the old Mother Goddess – if she was there at all – was seen no longer as life-giving but as dark and chaotic, and had to be slain for the sake of light and order. The Babylonian god Marduk slays the Sumerian Mother Goddess Tia-mat, depicted as a dragon, by splitting her dead body in two to recreate Heaven and Earth.

The story of Adam and Eve in the Garden, which stands at the beginning of our Western tradition, registers this disruption of the original human bond with Earth. Yahweh-Elohim – creating the world apart from himself – brings a new polar-ity into life, of transcendence and immanence. In one of three creation myths in Genesis – when taken literally – we inherit a transcendent God, a fallen universe, original sin inherent in human nature, a curse on the Earth and child-bearing, and, more pervasively perhaps, a distrust of our own spontaneous natures from which Imagination comes. Numinosity was now found in what could *not* be seen or touched, the 'graven image' was forbidden, and even the longings of the heart were to be punished:

> If I beheld the sun when it shined, or the moon walking *in* brightness; And my heart hath been secretly enticed, or my mouth hath kissed my hand. This

also were an iniquity to be punished by the judge: for I should have denied the God that is above.

(Book of Job 31:26–8; OUP, 2008, p. 444)

Consciousness, for the first time, is defined as wholly beyond creation, not immanent in, or as, creation.

This worshipping of Transcendence was inherited by the Roman Christian Church – in astonishing contrast to the immanence of the unedited Gnostic texts of the teachings of Jesus, only rediscovered in an urn at Nag Hammadi in Egypt in 1948. In the Gospel of Thomas, we read:

Logion 77: Jesus said: 'I am the All, the All came forth from Me and the All attained to Me.

Cleave a piece of wood, I am there; Lift up the stone and you will find Me there.'

(Guillamont, 1959, p. 43)

Logion 113: His disciples said to Him: "When will the Kingdom come"?

Jesus said: 'It will not come by expectation; they will not say "See, here", or "See, there"'.

But the Kingdom of the Father is spread upon the Earth and men do not see it.

(ibid., p. 57)

By the 4th century AD, the orthodox Roman interpretation of Christianity had completely suppressed the Gnostics' understanding of a Jesus who belonged to this Earth, proposing instead a Heaven for believers and a Hell to which 'pagans' went, a name from the Greek, '*paganos*,' meaning 'of the countryside.' One seal remains (Figure 7.1).

Jesus hangs outstretched on the cross, crucified, as one with *Orpheos-Bakkikos*, with the crescent moon of rebirth at the crown of the cross, and the seven stars of the Pleiades arching above, also known as the Lyre of Orpheus. Orpheus and 'Bacchus/Dionysus' were also Greek names for Osiris, placing Jesus in the 3,000-year-old immanent tradition of the dying and resurrected gods of the Earth's annual cycle of renewal and transformation. The image invites us to see through the outward names and forms to the shared essence within, so all these manifestations of the sacred are enriched through their mutual affinity – a joyous way of seeing life called Syncretism. But though the Gnostics' Jesus belongs to the millennial tradition of a divine Earth, orthodox Christianity took 'Christ' out of the seasonal cycles of natural life, proposing him instead as the saviour of human history from the ceaseless round of an inanimate Earth. Doctrinally, then, when taken literally, Christ would 'return' only at the end of time upon which time would end, and the world would then be 'redeemed' – though from what?

Figure 7.1 Jesus as Orpheus and Bacchus/Dionysus, Haematite cylinder seal, c. 300 AD; formerly in the Berlin Museum (now lost)

Source: from Joseph Campbell (2013, p. 255); used with permission

For nearly 2,000 years, formal Christian doctrine has perpetuated the oppositional paradigm of a Soul or Spirit belonging exclusively to human beings, while Earth and all Earth's non-human creatures were deprived of Soul or Spirit, and so presumed to have no feelings or needs, no intrinsic life of their own. They became simply inanimate objects to serve human purposes, effectively dead matter, forgetting that the word 'matter' came originally via *mater* for mother. This can be seen in the history of the Church's attitudes to women and childbirth – the 'faeces and urine of birth,' as St. Augustine puts it (Baring and Cashford, 1993, p. 530) – as well as anything 'earthly' belonging to 'this world,' along with its focus on original sin and the corresponding need for redeeming it. The idea of the universe as consciousness immanent in all creation was from the 11th century called 'heresy,' a word ironically coming from the Greek *hairesis*, to choose. Over the centuries, the Papal Bulls insisted on the literal 'virginity' of Mary, and also (logically) of her lineage, but did eventually concede to millions of pleas from Catholics all over the world for Mary to play a more significant role in the Christian story. She was declared 'Assumed into Heaven – Body and Soul' in 1950, and 'Queen of Heaven' in 1954 – to tumultuous applause in St. Peter's Square in Rome. But in a still doctrinally fallen universe, she was never called Queen of Earth. In many paintings and sculptures, she sits enthroned with the moon beneath her feet, often drawn as the black dragon of death complete with claws and tail, and Earth continues to be excluded from any kind of divinity (ibid., pp. 547–608). This polarisation is typically justified by a radical distinction between 'Spirit' and 'Nature,' which (to make matters worse) is not confined to doctrinal Christian thinking – formally arising with the pre-Socratic Greeks, and still lurking within the 18th century Enlightenment's 'worship' of Reason over Passion, and the superiority of the Inner over the Outer world.

Spirit and nature

When 'Spirit' is contrasted with, or more doctrinally, *opposed to*, 'Nature,' it carries the implicit assumption that this is the way things are – that is, an objective reality; sad but true! Yet oppositions inevitably arise together, the result of a new way of looking at what was once one whole. It is then surprising to discover that the origin of our word 'Nature' (deriving from Latin) as an oppositional term to 'Spirit,' was only formulated by the pre-Socratics in Greece around 600 BC. Although people in Original Participation lived in a numinous world, this world would not have been called 'Nature' in the way in which we now use the term. 'Nature' is a Latin word from *Natus* meaning being born, and is itself a translation of the Greek *Phusis* – a verbal noun based on *phuo*, to 'grow, appear.' C.S. Lewis, Barfield's great friend – one of the 'Oxford Inklings,' along with J.R.R. Tolkien – points out, in his book *The Discarded Image* (1964), that *phusis* was an idea first invented by the pre-Socratics around 600 BC to talk about the great variety of growing phenomena under a single name – *phusis* – instead of each place having its own different individual name for their own kind of growing phenomena, and everyone getting in a muddle. The specific particularities of what grew where, and what did not, had become less important than the convention of understanding the idea common to all of them.

However, Aristotle (384–322 BC) later divided the universe into two regions: the lower region of change he called *phusis*, literally now 'growth' (from which we get *physical and physics*), and the upper region he called *ouranos*, sky (the name of the god *Ouranos*). But because the sky also 'changed' with atmosphere and weather, he proposed a still higher level above sky, which he called Aether, which did not change and so was where the divine unchanging gods lived. By introducing this religious element, and confining it to what was both changeless and invisible, the implication followed that everything 'below the moon' – that is, where *Ouranos* ended and *Phusis* began – was not divine. Having opposed eternity and time, 'change' inevitably took on the ideas of irregularity and inconstancy – disregarding the changeless pattern of phases of change in the lunar model of waxing and waning – corresponding to the living patterns of growth, death, and rebirth taking place on Earth below. Cicero took this up, translating *Phusis* into *Natura* from *Natus*, birth, which was now contrasted to death (whereas growth, flowering, death, and rebirth were all originally understood as inherent phases of the divine process). Along with Aristotle, Cicero placed all that is eternal 'Above the Moon.' So by Roman times, *Natura* was nearer to a concept than an original deity, and the former identification with Mother Earth, Goddess of All, was lost, and Earth moved still further away from the hearts of human beings (Cashford, 2016, pp. 178–81). Lewis comments:

> Nature may be the oldest of things, but *Natura* is the youngest of divinities ... "Mother Nature" is a conscious metaphor. "Mother" Earth is something quite different. When, in Greek times, for instance, she lies beneath Father Sky, he begets, she bears. You can see it happening. This is geniuine mythopoeia. But while the mind is working on that level, what, in heaven's name, is Nature? Where is she? Who has seen her? What does she do?
>
> (Lewis, 1964, pp. 37–8)

It may be significant that, while many people call our Earth 'She,' the usual unreflective pronoun for 'Nature' is 'It.' This essentially depersonalised language is then extended to all creatures – invariably calling an animal or bird 'it' rather than he or she (even when the gender is obvious), and describing them in clauses beginning with 'which' and 'that,' rather than 'who.' More disturbingly, this distinction is enshrined and resolutely defended in official editing books on grammar, which declare anything else to be 'wrong.' Only our 'pets' get away with personhood as an indulgent exception of our own! In such minutiae as this is the old paradigm of Earth as a spirit-less, soul-less resource for the human project continually reinforced. So, there is a loss of presence buried in our everyday language, which those of us who see both 'Earth' and 'Nature' as sacred, may underestimate. It follows that any language which is anthropocentric, manipulative (using Earth as a resource), oppositional, dualistic, divisive, hierarchical, spirit- and soul-less belongs to the 'withdrawal stage' of participation, and so persistently inhibits the imagining of a new kind of relationship with Earth.

The internalisation of the outer world

The Withdrawal of Participation intensified in the mid-17th century in the West, when what was left of Earth's wisdom was transferred to humans, and specifically the saving power, so it was thought, of Reason in the Age of Enlightenment. Human consciousness could now expand inwards and learn to name and control those outward phenomena whose beauty would otherwise tempt it away from the moral law and the law of Reason (believed by many at the time to be the same). The further loss of numinous power in what could be seen and felt in the outer world of Nature left humanity more free to shape its surroundings – though also more alone in the midst of them and so less open to their correction when things went wrong. Nonetheless, the freeing of the outer and inner worlds from each other would allow each to be explored separately, making possible empirical science at one pole and a psychology of consciousness at the other, as well as all stations in between. Out of this latest phase, the flowering of the rational mind was fostered which was supposed to defy the once overpowering numinosity of Nature and banish what it now called superstition.

Owen Barfield defines this further stage as 'Internalisation' (Barfield, 1967, p. 171), where the inner world is emphatically opposed to the outer world. We can see it in the proliferation of words expressing something new in human thinking, and *about* human thinking. John Locke, for instance, in 1634, adopts the word already coined – 'consciousness' – but defines it for the first time as a 'perception of what passes in a man's own mind' (Locke, 1961, vol. 1, p. 87; Barfield, 1967, p. 171), giving *self*-consciousness its modern meaning. That was ten years before Descartes came up with his *Cogito ergo sum* (Miller, 1982, trans., *passim*). Since Descartes, philosophy has typically worked outwards from the individual thinking self, rather than inwards from the cosmos to the soul – as in Original Participation, and as Indigenous people still do. What T.S. Eliot called a 'dissociation of sensibility' (Eliot, 1932, p. 288), the poet Yeats memorably portrayed as 'that morning when Descartes discovered that he could think better in his bed than out of it' (Yeats, 1935, p. 192). Yeats also wrote of a 'bursting into fragments' (ibid., p. 189), the chief of which was a fragmentation of human beings from Nature, both outer Nature and our own inner nature, assumed up till then to be of the same essence.

Barfield warns this new kind of consciousness has become so invisibly embedded in the way we now perceive we can hardly recognise it:

> With the 17th century the consciousness of "myself" and the distinction between "my-self" and all other selves, the antithesis between "myself," the observer – and the external world, the observed – is such an obvious and early fact of experience to every one of us, such a fundamental starting point of our life as conscious beings, that it really requires a sort of *training of the imagination* to be able to conceive of any different kind of consciousness.
>
> (Barfield, 1967, pp. 164–71)

This new radical distinction between inner and outer kept on growing until it became 'self' and 'not-self,' and then 'self' and 'other.' And at this time, many self-reflective words suddenly appeared – 'self-knowledge,' for instance – which are now almost impossible to think without. But this division of self and not-self may have encouraged a *habit* of thinking in oppositions as a way of searching for the truth of anything, in direct contrast to the Imagination which seeks to unify opposites into a new whole.

The conception of 'laws' governing the outer world arises and grows steadily more impersonal, and then even what is already called 'matter' is described objectively and disinterestedly, while at the same time Earth ceases to be the centre around which the cosmos revolve. It was as though the European drive was to disentangle itself from what it now called its 'environment' – a word from the French, *environs*, which simply means 'round about' – while at the same time inevitably becoming more of a spectator than an actor. Barfield concludes that: 'We have to accustom ourselves thoroughly to the thought that the dualism, *objective, subjective*, is fundamental neither psychologically, historically, not philosophically' (Barfield, 2010, p. 204). It is, then, simply a stage in our thinking, a necessary one, but not inherent in the nature of things. Rather, it is the apotheosis of 2,000 years of polarisation, resulting in beliefs about 'the nature of reality' which are difficult to see, let alone suspend. This is what the literary critic Northrop Frye, in the introduction to his book *The Great Code: The Bible and Literature*, has called our 'mythological conditioning:' (Frye, 1982, p. xviii). The term suggests we may be caught in an old myth without realising it, and this prevents us from responding to Earth's cry for our help.

In the light of these comments, it is possible to see Jung experiencing and foreseeing the end of this 2,000-year period of history. When Jung writes that 'hemmed around by rationalistic walls we are cut off from the eternity of nature' (Jung, CW 8, para. 739), he is expressing his awareness of the limitation of Reason upon us, but almost as if this is inevitable. However, in his *Autobiography*, it is as though the 'eternity' of nature has come irresistibly to life, when he describes his feeling of being 'spread out over the landscape and inside things, and am myself living in every tree, in the splashing of the waves, in the clouds and the animals that come and go, in the processions of the seasons' (Jung, 1965, pp. 225–6). And it is significant that some pages later he says: 'The more uncertain I have felt about myself, the more there has grown in me a feeling of kinship with all things (ibid., p. 392). The implication here is that it is *uncertainty* which opens the heart and releases us from the dualism which both separates us from Earth and interprets the life of Earth as 'nothing but' our own 'projected' life. Similarly, Coleridge's suggestion of 'a willing suspension of disbelief for the moment' (Coleridge, 1975, pp. 168–9), and John Keats's plea for 'Negative Capability,' that is, when a man is capable of being in 'uncertainties, mysteries, doubts, without any irritable reaching after fact and reason' (Keats, 1952, p. 71), also summon the virtue of 'not knowing' to free the Imagination.

Final participation

To recapitulate: Final Participation is the opportunity to bring back together the human soul and the Soul of the World at a new level through the Imagination – by which is meant not only all the Imaginative Arts, but also bringing Imagination into all aspects of life. Einstein often said that 'Imagination is more important than knowledge. For knowledge is limited, whereas imagination embraces the entire world, stimulating progress, giving birth to evolution' (Calaprice, 2000, p. 10). Since myths are also forms of Imagination, a new way of being inevitably seeks a new myth or, equally, to turn it around, a new myth, becoming numinous, changes our way of being in the world. As Joseph Campbell said:

> The old gods are dead or dying, and people everywhere are searching, asking: "What is the new mythology to be, the mythology of this unified earth as of one harmonious being?"
>
> (Campbell, 1986, p. 17)

This, in Barfield's third stage of Final Participation, would be a mythology which is held both with conscious awareness and with the sustained courage and passion of Imagination, allowing us to experience our essential identity with Earth in a 'new way' – one both more modest and responsible. Restoring poetry to our vision resembles Thomas Mann's understanding of restoring myth as a 'late and mature' stage of the individual which comes out of the early youthful engagement with myth, but renders it conscious, and so can live it and not be lived by it (Mann, 1936, pp. 89–90). This union, or reunion, with our sacred world would be in mythological terms the reunion of the goddess and the god in the 'child' of a new vision beyond them both.

In one sense, the only change which is needed – if 'only' were not in this case 'everything' – is for us to experience our Earth again as '*Thou*' – as many people already do without even thinking of it. Yet this is not 'the same' *Thou* of some thousands of years ago, but one unique to our time – a new '*Thou*' – arising from a consciously willed choice which we, or our Imaginations, have made for the first time. The difference between the two states of mind spans the evolution of consciousness from Original Participation, through Withdrawal of Participation, to Final Participation. Here, the world becomes again a *Thou*, but a *Thou* with all the complexity of any personal relationship. Our Inner Nature would be free to unite and become one with Outer Nature – both experienced as aspects of each other – but now with the awareness inherited from the last evolutionary phase. We now know that we ourselves have brought about the crisis in which the inherent life of Earth now depends upon the kind of relationship we are able to make with Her. The question then becomes: can we transform ourselves into willing and compassionate participants in the unfolding of Earth's Mysteries which, for all our so-called intelligence, we barely even understand?

Barfield's insights draw on Shakespeare and the Romantic movement in Germany and Britain, especially Samuel Taylor Coleridge, whose work on Imagination came more and more to include the idea of the conscious will, without 'harming' Imagination or driving it away, but on the contrary, bringing Imagination into the realm of a higher consciousness as a 'unifying, synthetic power, which brings the whole soul of man into activity.' He writes: 'Imagination is the soul that is everywhere, and in each; and forms all into one graceful and intelligent whole.' It reveals 'the eternal in and through the temporal' (Coleridge, 1975, p. 174). If we look with William Blake's 'double vision' – with the inner and the outer eye together – then nature and human nature become permeable to each other:

> A double vision my eyes do see,
> And a double vision is always with me.
> With my inner eye 'tis an old man grey,
> With my outer a thistle across my way.
> (Keynes, 1961, p. 860)

'To a Man of Imagination, Nature is Imagination itself,' Blake says, 'As a Man is, so he sees.' (ibid., p. 835). So here, it is we who, following Blake's insight, have to move beyond our customary ways of being and seeing in such a way as to transform them into a new whole – one where Nature may finally be seen anew as 'Imagination itself.'

Thomas Berry: the great work

If this sounds unduly idealistic, let us listen to the words of Thomas Berry (1914–2009), a lifelong Passionist Monk, who would introduce himself with a twinkle in his eye as someone who at the age of 11 fell in love with a meadow and decided then and forever afterwards that 'Good is what is Good for the Meadow.' Earth comes first, and humans cannot be moral beings without a corresponding morality to Earth. All his studies opened out into the rest of the world. His study of theology became a study of history, Western history became Asian history and the history of Indigenous Peoples, and human history became the history of the Earth – past, present, and future. Whatever was missing from a particular dimension of thought, *that* was what he sought to explore, forever reaching for the widest, most comprehensive, whole: the Story of the Universe (Berry, 1999, *passim*). He was, then, in a unique position to place human and Earth history within the Universe Story, and to understand, as Mary Jayne Rust also points out (Finding the eye of the storm: see Chapter 2), we are now 'between stories' (Berry, 1978, pp. 77–8). Like Barfield, Thomas diagnosed a way out of our current impasse through the Imagination: 'Loss of Imagination and loss of Nature, they're the same thing' (Telephone conversation, 2008). The way to imagine the New Story and bring it to life was to include *all* the beings in the universe in their own right, given to them with their birth – their own particular right to be who they are: whether predator or prey, mountain or valley, river, sea or forest, birds, fish, insects, trees, flowers – animal, vegetable, and mineral. In his Schumacher Lectures and in his last books, he taught and wrote the manifesto *Every Being Has Rights*. He explored the idea that

> the natural rights of natural beings come from the same source as human rights: from the universe that brought us, that brought all things, into being ... the right to be, the right to habitat, and the right to fulfill one's role in the great community of existence.
>
> (Berry, 2006, p. 149)

It followed from this that existing laws oriented only to human beings had to be transformed to include the whole Earth community of which humans are only a part. He called this Earth Jurisprudence. By now he was calling himself not a Theologian, but a *Geo*logian, pointing to the 'grand liturgy of the universe' as sacred in itself, beyond any and all categories of faith. This is what he says:

> The natural world on the planet Earth gets its rights from the same source that humans get their rights, from the universe that brought them into being. . . . Every component of the Earth community has three rights: the right to be, the right to habitat, and the right and responsibility to fulfil its role in the ever-renewing processes of the Earth community. All rights are species-specific and limited. Rivers have river rights. Birds have bird rights. Insects have insect rights. Humans have human rights. Difference in rights is qualitative, not quantitative. The rights of an insect would be of no value to a tree or a fish. Human rights do not cancel out the rights of other modes of being to exist in their natural state. Human property rights are not absolute. . . . Each component of the Earth community is immediately or mediately dependent on every other member of the community for the nourishment and assistance it needs for its own survival. This mutual nourishment, which includes the predator-prey relationship, is integral with the role that each component of the Earth has within the comprehensive community of existence.
>
> (Berry, 2006, pp. 149–50)

Thomas was not entirely happy with the language of rights, but it was the best we had to be going on with. Since, in 1886, the collective lives of corporations were, astonishingly, given the rights of individuals, then so at least should Earth, in all her gloriously diverse manifestations, have the rights of individual life. But we have to begin from where we find ourselves. So law is necessary where morality has failed, and morality is necessary where love is absent. If you know and love the meadows and the woodlands and the rivers and all the creatures who live in their community, then you feel *with* them and understand what they need to flourish. Their wounds become our wounds: we are all mutually dependent, mutually reflecting and mutually enhancing. That is why we need to learn the language of mountains and rivers, trees, birds, animals, and insects, and also the language of the stars in the heavens.

Is this not – in one of its infinite dimensions – what Final Participation might look like? If, with this in mind, we return to the protest of the Greek Earth at the unwitting crimes of Oedipus, then it seems quite reasonable that king and Earth would be bound together in mutual relationship, and that the Earth would suffer the king's folly, both in itself, and also as the only way to bring the king to an awareness of what he has done, and who he is. It is perhaps no more than what is happening now all over the planet, only we have learned for many centuries to ignore or explain away the connection between human acts and natural devastation, habitually reinterpreting the causes and effects so we can rarely recognise them as our own. Gaia called out once, and She was heard. But what if, in a Silent Spring, Gaia calls, Persephone calls, and no-one is listening? (and see chapter 13, this volume) And now there is no Delphic Oracle to remind us of who we are and what is true, unless we seek it in our hearts and trust what our hearts say to us. So, to end with Einstein:

> Human beings are part of the whole called by us "the universe," a part limited in time and space. We experience ourselves, our thoughts and feelings, as something separate from the rest – a kind of optical illusion of our consciousness. This delusion is a kind of prison for us, restricting us to our personal desires and affection for a few persons nearest to us. Our task must be to free us from this prison by widening our circle of understanding and compassion to embrace all living creatures and the whole of nature in its beauty.
>
> (Calaprice, 2000, p. 316)

References

Barfield, O. (1967). *History in English words*. Edinburgh: Floris Books.

———. (1988). *Saving the appearances: A study in idolatry*, 2nd ed. Hanover, NH: Wesleyan University Press.

———. (2010). *Poetic diction: A study in meaning*. Oxford: Barfield Press.

Baring, A. and Cashford, J. (1993). *The myth of the goddess: Evolution of an image*. London: Penguin Books.

Berry, T. (1978). The new story: Comments on the origin, identification and transmission of values. *Teilhard Studies*, 1 (Winter).

———. (1999). *The great work: Our way into the future*. New York: Bell Tower Books.

———. (2006). *Evening thoughts: Reflecting on earth as a sacred community*, ed. by M.E. Tucker. San Francisco, CA: Sierra Club Books.

Book of Job. *King James bible* 16 (2008). Authorized King James Version. Oxford University Press.

Calaprice, A., ed. (2000). *The expanded quotable Einstein*. Princeton, NJ: Princeton University Press.

Campbell, J. (1986). *The inner reaches of outer space: Metaphor as myth and religion*. Novato, CA: New World Library.

———. (2013). *Goddesses: Mysteries of the feminine divine*. Novato, CA: New World Library, p. 255.

Cashford, J., trans. (2003). *The Homeric hymns*. London: Penguin Classics.

———. (2016). *The moon: Symbol of transformation*. Carteton, Oxfordshire: Greystones Press.

Coleridge, S.T. (1975). *Biographia litteraria*, Ch. IV. London: J.M. Dent.

Eliot, T.S. (1932). *Selected essays*. London: Faber & Faber Ltd.

Frye, N. (1982). *The great code: The bible and literature*. New York and London: Harcourt, Brace and Jovanovitch.

Grene, D. and Lattimore, R., trans. (1960). *Greek tragedies, vol. 1. Sophocles' oedipus rex*, 2nd ed. Chicago, IL: University of Chicago Press.

Guillamont, A. (1959). *The gospel according to Thomas*, Coptic text established and trans. by A. Guillamont, et al. Leiden: E.J. Brill and London: Collins.

Jung, C.G. (1953–77). *Except where indicated, references are by volume and paragraph number to the collected works of C. G. Jung*. 20 vol., ed. by H. Read, M. Fordham, and G. Adler, trans. by R.F.C. Hull. London: Routledge and Princeton: Princeton University Press.

———. (1959). Face to face. Interview with John Freeman. BBC.

———. (1965). *Memories, dreams, reflections*, recorded and ed. by A. Jaffe. London: Fontana Paperbacks, Flamingo Edition.

———. (1973). *Letter to P.W. Martin (20 August 1945). C.G. Jung Letters*, selected and ed. by G. Adler in collaboration with A. Jaffe. London: Routledge and Kegan Paul.

Keats, J. (1952). *Letters of John Keats*, ed. by M.B. Forman. 4th ed. Oxford: Oxford University Press.

Keynes, G., ed. (1961). *Blake: Complete poetry and prose*. London: Nonesuch Press, p. 835.

Lewis, C.S. (1964). *The discarded image: An introduction to medieval and renaissance literature*. Cambridge: Cambridge University Press.

Locke, J. (1961). *An essay concerning human understanding*, 5th ed., ed. by J.W. Yolton. London: Dent (first published 1706).

Mann, T. (1936). *Freud and the future. Life and Letters Today*, trans. by H.T. Lowe-Porter, 15, pp. 89–90. Reprinted in *Freud, goethe, wagner* (essays), trans. by H.T. Lowe-Porter and R. Matthias-Reil. New York: Alfred A. Knopf.

Miller, R.P. (1982). *René Descartes, Principles of philosophy*. Translated, with explanatory notes. New York, NY: Springer Books.

Shelley, P.B. (1995). A defence of poetry. In *Poems and prose*. London: J. M. Dent.

Yeats, W.B. (1935). *Autobiographies*. London and Basingstoke: The Macmillan Press.

Ruin of the Lesser Kestrels

So many mysteries in this land
How is life sustained?
There are no separate identities;
No edges
The sun claims everything
Including the brain's
Hard won gift of discrimination
Entities flow together
Softened on the anvil
Of the plain,
To be hammered
By the sun's imagination
Into now

                    Grant Clifford

Part III

Symbols of transformation

Chapter 8

Engaging the Green Man, breaking our spell of enchantment

Jeffrey T. Kiehl

Beginnings

A reluctance in writing . . . I feel an impediment of great intensity when contemplating writing this chapter. Thoughts circle, ideas appear and disappear. There is much fluidity, but little concretion. Nothing is congealing and coming to form. Here is where the work begins. It is the old alchemical process of *solve et coagula*, a state of dissolution yearning to come together, to take on form. How will this work begin? How will the words appear? I don't know at this stage. All I know is I must begin to write and continue writing to see what happens. As I reflect on the purpose of this book, I feel myself enter a state of reverie – and just at this moment, the Green Man comes to me. He dwells in me and I sense his presence in the surrounding more-than-human-world. There is a sense of excitement with this image, it becomes the emergent theme. I open myself up to this possibility.

The Green Man and enchantment

As often happens, after discovering a theme, I begin to read as much as possible on the topic. I already had several books on the Green Man, some purchased thirty years ago when I lived in Cambridge, England. Now, after all those years, I seek out more books on 'The Green Man.' As I begin to read, I feel the Green Man distancing himself from me. When I stick with the images in the Green Man books, he is very close, but the moment I begin to read about him, he drifts away.

In this moment, I realize I cannot write this chapter from an academic perspective. Furthermore, I realize it is just this perspective which has driven the Green Man from our post-modern collective consciousness. He receded from us because we became too pre-occupied with a particularly narrow world: a highly linear, rational, thinking worldview. So, if I am to engage with the Green Man, I need to forgo any approach laden with rationality.

The other image I become aware of is enchantment and how it relates to our current state of catastrophic climate crisis. I come to this awareness when re-reading Joseph Campbell's retelling of the Parzival story (1968, p. 492), an old story full of magic and enchantment. I am struck by the story of Gawain's encounter with the Castle of Marvels. The story of Gawain's encounter with the Green Knight also comes to mind (Tolkien, 1975). These mythic breadcrumbs lead me to other stories of enchantment from Celtic mythology to on to fairy tales involving the casting of magical spells, perhaps the most emblematic being Sleeping Beauty. A multitude of images well up, indicative of the richness of imagination and how it plays with one's psyche. So, psyche has gifted me with the images of both the Green Man and enchantment. What to do? How would I, how could I, weave these images together? What are the relationships of these images to our current state of climate collapse, climate crisis, climate disruption?

We can no longer escape the trauma of climatic disruption. With each new story of ecological collapse, our sense of time collapses. Changes predicted far in the future occur before our eyes. How can we imagine our way through these difficult times? Is there a way, or are we doomed to sink into a state of destructive darkness? We have reached a point where people do not know what to do. Jung believed in such times we should perhaps ask the unconscious for help (CW 18, para. 599). We can no longer rely on our limited conscious abilities to sort out a problem as immense as climate disruption. We need to peer into the unconscious and seek advice from this depthless realm of psyche. Dreams and engagement with the imaginal are powerful ways to listen to what the unconscious tells us. What wisdom lies within the unconscious about dealing with climate disruption?

If we enter the realm of the imaginal, what boon could be brought back to the day world of our great suffering? Specifically, is it possible the Green Man is calling to us? What happens if we turned our attention to him and listened to what he tells us? In this moment, another form appears: she is Sheela-Na-Gig – the primordial feminine who exposes herself to us in full awareness, the Green Man's feminine complement. We need to re-engage with the magic, mythic dimensions of these archetypal forms for their presence in our lives enriches and ensouls us.

Playing with these images allows us to see connections between our current climate catastrophe and the imaginal forms of the Green Man and enchantment. The Green Man is an ancient, rich image of Nature. Just gazing at his form evokes a sense of primordial energy associated with the more-than-human world. The piercing gaze of his eyes, the leafy covering over much of this face, and the vines flowing from his open mouth are vivid reminders that we *are* Nature. His face emerges from wood and stone reminding us we too are emergent beings on this living, dynamic world (see Chapter 12 in this volume). Collectively, we have forgotten the Green Man. He is relegated to history, tucked away in books. The closest contact many have with his image is a chance encounter when visiting an old village church or cathedral. Yet, his form is ever present.

Enchantment? We are all in this state. For many still do not consciously recognize what is happening to the planet or purposively choose not to recognize. How can we break this spell, this enchantment, which prevents so many from action? How can we reconnect to the animate spirit of the Green Man? How can we invoke imagination in this time of collective enchantment? These are the questions I explore in this piece. Note – a central word to these questions is 'imagine.' Following this intuitive imagistic hint from psyche I let my imagination lead in answering these questions.

Our current enchantment manifests in many ways. One is in a complete detachment from what is happening to the environment. Our acceleration towards global ecocide goes unnoticed by far too many. We continue to consume in a state of complete ignorance, given the finite resources available. People remain mute about these disturbing realities. There are also those who reject the science and willfully do everything possible to distort the science of climate disruption. Despair can also be a form of enchantment. We can become so immersed in our feelings of anger and grief we become paralyzed; we are literally frozen. We may also become enchanted by technology, for it offers us ways to 'geo-engineer' our way out of warming. It promises solutions which are extremely seductive, for enchantment also appears in the form of a seduction. All of these and more are ways we can be enchanted into ignoring the immediacy of the climate crisis. They all separate us from our deep connectedness to Nature and lead to a disengagement with the world.

Psychologically, it is unhealthy to relate to the world from an overly one-sided viewpoint. Jung felt one-sidedness was the root of neurosis, our ill adapted relationship with the world in which we live (CW 13, para. 15). When we invest so much psychic energy into a conscious one-sided perspective, we lose critically essential aspects of ourselves. This loss can happen individually, and also collectively. Cultures can be overly one-sided, too! We are currently witnessing such a phenomenon not just with the climate crisis, but with immigration and nationalism. Of course, rejected aspects of ourselves are not completely lost. From a depth psychological perspective, what is consciously rejected falls into the unconscious. From a Jungian perspective, the unconscious not only holds that rejected from the outer world but is also populated by archetypes; imagistic patterns of instincts that were never repressed. Jung further noted that archetypal patterns are rooted in Nature itself (CW 10, para. 53). By connecting to archetypes, we experience the numinosity or Nature. As stated, problems arise when one-sidedness cuts us of from this experience. Psyche senses this imbalance and acts to rebalance our relationships, inner and outer. The self-regulating dynamics of psyche involve the activation of archetypal patterns whenever we drift too far from wholeness. This holistic dynamic model of psyche is one of Jung's most important contributions to psychology. It tells us psyche 'self-corrects' our collective one-sidedness through the appearance of archetypal images.

If this is the case, then how is the collective unconscious reacting to our current, state of environmentally destructive one-sidedness? The global climate crisis is a result of our two-hundred-year reliance on fossil fuels, which arises from our insatiable desire for energy. Over this time, we used our highly skilled rational sensate thinking functions to develop technologies to generate this energy. If rationality is the collectively dominant conscious function, then feeling/intuition must have fallen into the unconscious. The pathologic expressions of our lack of feeling/

intuition are expressed through the current states of collective dissociation, impulsivity and compulsivity.

Note that for Jung, the word 'feeling' is best interpreted as 'valuing.' This interpretation resonates with what takes place in fairy tales, in which the ignored feminine casts a spell on the community as punishment for their dismissive behavior towards her (see Chapter 14 of this volume). When the collective is possessed in this way, it cannot recognize, value, or understand what is going on in the world. The spell of enchantment blinds us to our actions. Often, we can break the spell by honoring the dismissed, rejected feminine. We can awaken the feminine (Sleeping Beauty) by putting ourselves in a sensuous relationship with her. What state of consciousness is required to break the spell? What are the qualities of Parzival, Gawain, or the prince of fairy tales which breaks the spell? Of course, not all spell breakers are male. The individual who break spells is one who is heart-centered, rather than power-centered. In the tales, these individuals are closer to Nature and the feminine, and often receive help from Nature to carry out their task. These tales provide us with the archetypal patterns required to break spells. If we accept that our current situation concerning climate is one of enchantment, then the tales of over one thousand years ago still speak to us of how to avoid the worst consequences of climate catastrophe.

What of the Green Man? It is interesting that when the Arthurian tales were arising in Europe, so too were the images of the Green Man and Sheela-Na-Gig. Psychologically, the collective unconscious was reminding us of our deep connection to Nature and that we were beginning to separate from Nature. It is important to remember, at the time (12th century), the cult of the virgin Mary was appearing in Europe (Kiehl, 2016). So, archetypal forms of the masculine and the feminine were arising out of the unconscious to correct our growing one-sidedness. I feel these stories and images still carry tremendous numinosity and we can heed their call to address our current climate crisis. The stories of our falling into a collective sleep remind us what happens when we separate ourselves from Nature. By dismissing the feminine and Nature, especially the dark feminine, we invite being enchanted.

Another inherent motif in these old stories is of wounding. Often, the enchantment traces its origins back to some form of wounding: a lance to the thigh, a pricking of the finger, an invitation not given. As noted, this may include a transgression against the feminine. Currently, we are inflicting severe wounds on ourselves and Nature. The burning down of the Amazon, the burning of fossil fuels, the fouling of oceans and air are all explicit forms of wounding Nature.

It is important to recognize that we did not create the Green Man, or the Castle of Marvels, or the enchantment of Sleeping Beauty. These archetypal images were given to us from the otherworld. Tolkien stated he did not write the story of Treebeard in the *Lord of the Rings*, but he allowed the story to flow through him onto the page (Tolkien, 1981, note on pp. 211–2). When we read such stories, we immediately feel they are different from other forms of literature. We know these stories carry numinosity and a truth surpassing other forms of writing. Jung felt the source of all creativity is the play of the imagination (CW 6, par. 93).

In today's climate crisis, we need as much creativity as possible. We are not going to *think* our way out of this situation, for it is thinking that got us to this. In reflecting on the Green Man and tales of enchantment, we are immediately placed in the realm of the imagination. Psyche dwells in the realm of the imaginal; and the imaginal is the portal linking the phenomenal world of everyday events to the world outside of space and time, the world of 'Once upon a time.' This portal sits at the nexus of the local and non-local, the temporal and the atemporal. Myth and all its rich imagery arise from this other world. Imagination reconnects us to our primal, archaic beginnings, our source. It exists so we can remember our participation in Nature.

As noted, Jung pointed out that archetypes are rooted in earth itself. Thus, to play with archetypal images is to experience our rootedness in earth. As such, archetypal images and stories of the Green Man provide portals to a deep, numinous felt experience of earth. In the days of old, we were more receptive to communicating with earth's soul, *Anima Mundi*. We were open to the magical and mythical dimensions of perceiving the world. We could carve the image of Green Man into the Christian cosmology and be fine with it. We could even chisel in stone the feminine openness of Sheela-Na-Gig without offense. Nature was continually revealing herself to us, and we were receptive to her. The presence of the Green Man allowed us to transit between the sacred and the profane. In fact, there was little separation between sacred and profane. If the sacred is recognized as everywhere, then there is no place to stand (*pro-phanos*) outside of her. With the development of our hyper-rationality, we have become suspicious of Nature. What does she want from us? What am I going to get out of any relationship with her? Can I trust her?

We lost the ability to creatively hold the opposites inherent in Nature, the most fundamental dyad being Nature's duplex of Creation/Destruction. The motif of this primal dyad pervades the tales of old. The Green Knight is both terrifying and life-bestowing. The mother-witch grants and takes away at any moment. Nature holds fecundity and ferocity, and we are tasked with being able to sit with both. If we cannot hold this tension of opposites, then we fear Nature. In doing so, we unconsciously project our fear out onto Nature and devalue her. One way to work through our inability to dwell in a both/and world is through imagination.

Another important dyad exists in tales of old is of time and timelessness. For in a state of enchantment, time stops and we enter a realm of timelessness. Time plays a central role in life as clock for organic process. Beginning, growth, decay, death appear everywhere in Nature. The Green Man reminds us of this process, and he may be viewed as the consort of the Goddess (like Sheela-Na-Gig), who is a symbol of organic Nature. Thus, by being in touch with the Green Man, we are close to the Goddess, close to transiency. The more we forget this, the more we believe in our permanency on the planet. We believe we can do anything to

Nature; Nature is infinitely resilient, abundant, and able to serve a life of unbridled growth. In the tale of Gawain and the Green Knight, the knight reminds Gawain of the prescient dimension of life. The Green Knight is immortal, but Gawain is mortal. We too need to be reminded of our mortality which means we cannot act as if Nature is here to serve our desires.

Scientifically, time is very important to climate disruption. Science tells us we have only a few years left to get stop depending on fossil fuels before we commit to a warming exceeding 2°C. This sense of urgency is expressed in the Extinction Rebellion and Greta Thunberg's calls for immediate action. Time is accelerating and collapsing. Yet, everything we know about the imaginal, the realm of 'once upon a time' speaks of its atemporality. It does not exist in a realm of linear, chronological progression. Soon we will be 'out of time' from a scientific point of view, we feel 'time is up' to act 'in time.' When we enter the imaginal, we are in the time-less realm, from which things are perceived differently. Does the Green Man worry about what is happening in the world? Is he capable of knowing how much time is left? The Green Man, and his mythic friends, are ambiguous enough to take sides on this issue, which reminds me of the story in *The Lord of the Rings*, in which Pippin and Merry ask Treebeard whose side he is on. Treebeard is mystified by this question and responds by stating he is on no one's side. Yet in the end, he realizes he must take a side to save his sacred forest.

So, perhaps, one of our roles is to engage with the Green Man and awaken him to what is going on in the world. He needs to know how his ancient, verdant grounds are being decimated by human development and climatic chaos. Perhaps our current task is to do just this. If the Green Man became engaged in the fight for earth, then we would be tapping into an archetypal force of tremendous magnitude. The Green Man needs us in order to manifest in this temporal world. We need him to help us in our time of crisis. We are in reciprocal relationship with him. The crisis we are experiencing right now is a catalyst for accessing this archetypal energy. He can tell us how to break the spell of enchantment that has befallen much of humanity. He holds the keys to releasing us from our collective, dissociative somnambulant state.

How do we evoke the Green Man? My premise is the Green Man dwells in the realm of the imaginal. Thus, an academic exploration of the history of the Green Man, or a compilation of Green Man's appearance throughout the world, are not sufficient to experience his living reality. There are several excellent resources on the Green Man, and all provide a deepening experience of him. However, most of these sources are rooted in the realm of space and time, not the imaginal, which is where depth psychology, especially the approach of Jung, becomes important. He faced a critical moment in his life when he had to descend into the depths of psyche. His survival depended on his finding a new way to living life. The process he discovered to explore the realms of psyche, active imagination, is a powerful technique connecting us to the otherworld.

Active imagination begins by evoking an image or feeling into consciousness. You sit quietly holding the image and/or feeling in consciousness. As the image or feeling becomes more apparent (felt), you begin a dialogue with the image. You may ask the figure standing before you a specific question, or you may wait for the figure to engage with you. The initial encounter with a figure may take some time. However, experience shows once you have engaged with the image, the process takes little to no effort on your part. After (or even during) the process, it is important to concretize your experience through writing, artwork, or body movement. What follows is a series of active imaginations I carried out with the Green Man. I was transcribing our dialogue as we spoke with one another. Some of what takes place in our dialogue is difficult to understand for it transcends linear, rational thinking. I urge you, the reader, to reflect on the Green Man's messages. If would also encourage you to engage with the Green Man to discover what he wants to tell you.   .

Encountering the Green Man

Where can I find you, Green Man? Where do I need to go in order to encounter you? Ah, I see something forming before me . . . as I walk along this stone-lined, gravel path, I see an opening in the forest to my left. I feel pulled to leave this smooth path and wade through the knee-high brambles in order to enter the forest. As I take my first step into the forest, the light above me dims and I feel the coolness of the overarching tree canopy above my head. I enjoy this softer environment and notice the diminishment of the sounds of the outer world. The quietness of a forest is like no other. The path before me is strewn with fallen, decaying leaves and twigs that exude a fermented, fecund aroma. The alchemy of decomposing matter is strongly present here. I am reminded I tread on microscopic bacteria which are extremely important to earth's cycling of substances like carbon.

I pause and breathe in this organic-laden air, and then proceed along through the forest. Ahead I see an opening or clearing where the trees appear to have created their own inviting, circular mandala. The forest floor in this space contains a spiral of small smooth, white stones. At the center of the spiral is a large stone arch in which rests the carved face of the Green Man. I follow the spiraling path all the while feeling pulled between staring at the Green Man and looking down at the path. Eventually, I arrive at the center of the spiral and stand before the stone arch. I sense this arch is a remnant of an older, larger structure; perhaps a chapel, an ancient temple, or sacred burial site? I am not sure. The felt sense of this place holds tremendous numinosity.

I stand before the limestone edifice, the carved face resting in the keystone of the arch, just at eye level. I stare at Green Man with his moss- and vine-encrusted face. Before I can ask anything of him, he cries out, 'So, I see you have finally arrived!' I step back to see both rage and mirth in his face. His eyes meet mine.

The color of his face resembles to the leaves of the trees surrounding the arch, a shimmering light green. There is a faint earthy smell coming from his mouth:

I: What do you mean? How did you know I was coming?

GM: Your kind have been really screwing things up! I see it happening from this place. I hold up this arch as a testament to how things can exist in harmony. Humanity, however, has chosen to go its own way. A way not of ease, but of dis-ease!

I: How do we get back to your world?

GM: You are already in my world. You never left it. I am presenting myself to you all the time. I am before you all the time. You neglect to care for the world that presents itself.

I: Are you saying you are the cause of what is happening?

GM: No! You are the cause, but I carry the world's suffering as symptom. I hold both symptom and salve. I am both and neither.

I: What is this stone arch?

GM: The origins of this arch reach back to ancient times when man cared for earth, when human beings cared for the creatures of earth. The people who made this arch placed me at the center – the keystone – as a reminder of how central I was to life. I was here to watch, listen, and speak to humans whenever necessary. But over time, many years now, fewer and fewer chose to come see me, then one day people stopped coming altogether. Your people forgot I was ever here. Yet, the arch remains strong and still stands.

I: Yes, I have come to see you now because we do not know what to do. Our world is falling apart because of negligence and forgetting. Too few are awake to what is happening. They have fallen under a spell of enchantment. I have come to ask how this spell can be broken.

GM: I watched all of this happen. I can answer your questions. But you will not like what I have to say. You will not want to stare into the mirror I hold before you. You brought this enchantment upon yourselves! The more you valued purely rational beings, the more you forgot what it is like to be of Nature. You no longer walked in the forest, but began to measure and count the trees, rather than see their beauty. You rushed to measure everything: air, water, fire, earth. You reached a point where you could explain what was happening in the world, but you lost the ability to feel what was happening. Your academics, the so-called knowledgeable people, had only one way to know the world, unlike the ancients who knew ways your people no longer respect. With each step, you deepened your sleep. You forgot more and more. Nature became so far from you. You cut off your relationship with life, your own life. You became captivated by the things you produced, not created as the way of Nature. You wanted to manipulate the world, predict and control it. Everything had to have a 'use.' 'What's the use?' became your mantra. You also created a language which separates and dismembers your relationship to the world: 'I, me, mine.' Separate, separate, separate – you do it with your words, your grammar, your stories. Your fascination with language has also cast a spell on you for it narrows your way of seeing the world. You have also abandoned my sister, Sheela-Na-Gig. She was the creator who made all and was respected for her gifts to the world. Now she is forgotten and even worse despised. Rejecting her has wounded her deeply. How else do you enchant yourselves? By consuming everything around you. You feel empty inside and look around for ways to fill yourselves. So, you build things, sell things, buy things to assuage your hunger. You have even created a whole industry whose sole purpose is to seduce you into buying and consuming more and more. You are enchanted by the 'ten thousand things.'

I: Yes, but how do we break this spell of enchantment?

GM: Not an easy thing to do. If it were, someone would have done it a long time ago. If it were easy, then there would be far fewer tales of enchantment. To break a spell requires special knowledge, a talisman and helpers. You are living in an enchantment that defines the world you live in. How many are willing to question this? What sort of sacrifices are you willing to make to break the spell? What are you willing to let go of? I see the fear in your face as I ask these questions. The spell is deeply imbedded in your psyches. Your myths and religions perpetuate the spell. You are so enchanted that you do not want to leave that place. However, I see that you have come to me and I know there are others in your world who want to break the spell. Some are beginning to break through the illusions that keep them asleep. They catch a glimpse of what the world looks like outside of the enchantment. The disrupted earth is waking you up for you can no longer ignore her.

I: Yes, you are right. But I am here now asking for your help. How do we break the spell?

GM: I sense your honest desire to help. I am not heartless. I watched as my world was destroyed millennia ago, when the great forests were felled. My image was carved in this stone to remind everyone what life was like and how green the world was. I feel for you and those who want to wake up. I can only give you hints. Start by uncovering me. Look how hidden I am deep in a forest, vines and moss covering my face. I am invisible. Your task is to make me visible again.

I: What are the ways that I can make you visible? How can I get people to see you?

GM: I am always present. All people need to do is look more deeply at the world. They need to breathe me in, to taste me, touch me, smell me. They are able do all these things. Of course, there is this strange relationship you have with time . . .

I: What do you mean, 'this strange relationship . . . with time?'

GM: You think time flows along like a river in a channel, smoothly flowing along at each point. That is not how time is! Time is organic. Past, present, and future are always right here before you. Even your great scientist knew that, but few listened to him. You feel time is running out. I am timeless; yes, I live 'out of time' all the time! I am ever present. Wrap your mind around this. Let it flow through you. Feel what it is like to let go of your narrow, limited view of time.

I: You are avoiding answering my question about how to break the enchantment!

GM: Don't be a fool. Of course I am answering your question. Again, you are stuck in your little world and cannot experience the vastness. I am pointing you to something much deeper, much more profound. You would call it the *numinous*. This is the secret to breaking the spell of enchantment. If human beings could remember how sacred the world really is, then you would experience true fulfillment. There would be little desire to fill your minds and bellies with useless trash, which you then toss into the oceans, my oceans! Open yourselves senses, your psyches to seeing me as I truly am. For I am not just a carved face in stone. I live beyond this place. This is the deepest secrets of all: *your face and my face are the same*. The real spell is that you have forgotten who you are. I am your mirror!

I: Tell me, whenever I see images of you, there are vines extending out of your mouth. Why do these vines grow from your open mouth?

GM: What comes out of your mouth?

I: Words.

GM: These vines are my words. I have no control over these vines. They arise spontaneously from within me and seek outer expression. They extend out beyond me and entice people to look at me. Is this not what your words do? Do they not flow from a deep source within you? Do they not entice people to attend to you? Here I speak of words coming from the deepest part of your psyche. Not the words that are shallow, self-centered, but words of truth.

I: Yes, I agree. There are times when I just let the words come from a deeper place within myself. I observe them and how they affect those who are listening to them. Words can move us in profound ways.

GM: These are my vines. They remind people of how Nature can come out of you and connect over great distances. They enliven and enrich anyone who connects to them. I have revealed another way to break the spell of enchantment. Your voice, your words – if they arise from the depths of psyche, break the spell. Your poets and prophets know this.

I: Indeed, I realized that is why you move me. Your vines animate me.

GM: Yes, these types of words make the invisible, visible.

I: I feel an opening here. I want to thank you for showing me some ways to break through the enchantment.

GM: You are welcome.

One never knows what will happen during active imagination. The images and words arise from a deeper part of psyche, which realizes it is, in all humility, being asked to speak to us. I never know what is going to 'come up' in such dialogues. As I look back upon meeting the Green Man, there is a deep sense of sadness. I felt his anger, remorse, and frustration with our lack of concern about the living planet. Also, I have nostalgia for the old days, when he was an integral part of people's lives. He recognizes these days are past but yearns for the spirit they held for both him and the people. I empathized with his feelings of loss, for his loss is indeed ours, too.

I will not analyze what happened except for the part of our exchange concerning time. The Green Man lives in the eternal, an experience of being time-less, or 'out of time.' (see Chapter 11 of this volume). He could experience our relative sense of time, the feeling of things arising, living and dying, but this is not the world in which he dwells. His statement that past, present, and future are all here now and his reference to the great scientist refer to Einstein's 'block' universe. How is this relevant to our climate crisis? If we were to truly understand this reality, then we would realize that to wait makes no sense. Since the future is here now, we need to tend to it; in other words, we need to act for the future, now. Not 'as if' the future was here now, but to recognize that it is here now. I believe what the Green Man was telling me – and he was emphatic about it – is to embody the presence of all times here and now. The Green Man was doing this in the active imagination, and I was a part of his experience. There is also the sense we need to be aware and experience the state of timelessness. It opens us to seeing climate disruption from an entirely different perspective. Whenever I drop into such an experience through meditation, my felt sense of climate anxiety dissipates.

Concluding thoughts

Nowadays, I often experience a sense of pervasive hopelessness regarding the climate crisis. As a climate scientist, I worked on this for forty years and lived with the frustration of seeing the science being ignored and how little action takes place to get us off fossil fuels. As a Jungian analyst, I work with eco-anxiety, ever present in people's psyches. I listened to Jung's wise counsel concerning how to heal our split with the world; though even he at the end of his life experienced dark, pessimistic visions of our future (Kingsley, 2018, pp. 418–22). I reach a point where I must accept the psychic reality that we really do not know how things are going to unfold. I admit the unconscious holds things which are ineffable and transcend our relative experience of space and time. I am not using this as a 'cop out' from doing everything possible to avoid further climate catastrophe. However, I must admit I do not know. The Green Man, as well as Sheela-Na-Gig, holds such wisdom. We need to stay in touch with these archetypal images. Perhaps the Green Man can lead us towards breaking the spell of enchantment preventing us from collective climate action. We owe it to the Green Man – and to the planet – to break the spell that keeps us isolated and frozen. I believe that with the Green Man's help, we can do this.

References

Campbell, J. (1968). *The masks of god, creative mythology.* New York, NY: The Viking Press.

Jung, C.G. (1953–77). *Except where indicated, references are by volume and paragraph number to the collected works of C. G. Jung.* 20 vol., ed. by H. Read, M. Fordham, and G. Adler, trans. by R.F.C. Hull. London: Routledge and Princeton: Princeton University Press.

Kiehl, J. (2016). The evolution of archetypal forms in western civilization. *Psychological Perspectives*, 59(2), pp. 202–218.

Kingsley, P. (2018). *Catafalque, Carl Jung and the end of humanity*, Vol. 1. London: Catafalque Press.

Tolkien, J.R.R. (1975). *Sir Gawain and the green knight, pearl, and Sir Orfeo.* Boston: Houghton Mifflin Company.

———. (1981). *The letters of JRR Tolkien*, ed. by H. Carpenter. Boston: Houghton Mifflin Company.

Chapter 9

At war with the natural world

Nature as Other

John Colverson

The concept of nature has become reified by Western consciousness and has become a thing to be othered. To understand the meaning of other we can refer to Plato who, in *Meno* (88d), refers to '*alle psyche*' which is translated as 'the rest of the soul' (Barber, 1968). As Renos Papadopoulos, Professor of Analytical Psychology at Essex University, points out,

> this implies there is still some other part of the psyche which constitutes the rest of the one already mentioned. So, if the rest is added to the existing part, a wholeness, a totality may be achieved; but if the rest is left apart, then a separation, a division may result.
>
> (Papadopoulos, 1984, p. 56)

It is possible to draw parallels between the prejudice, disrespect, and objectification suffered by disabled people, minority communities, and non-human nature. While human groups have, belatedly, won some recognition, respect, and rights, non-human nature has largely not. The idea of non-human nature having rights is a foreign concept to most humans, and a laughable notion to many.

Children of the Mother Goddess

It is easy for us to believe that we are separate individuals, distinct from one another and the natural world – but this is not reality. We are as much part of the natural world as any living organism and might think of ourselves as a more recent experiment in nature's project. Yet we are seriously out of step with the natural world. We used to feel more connected to a sense of being part of something larger than ourselves: a living, breathing planet, which gives birth to an amazing array of life from the invisible microscopic world through mosses, fungi, and algae, plants and animals, to complex eco-systems with everything held in balance.

Many stone figures of the Earth Mother have been found throughout Europe, stretching as far as Siberia. It's thought these figures were made in the late Stone Age (Palaeolithic) period and represent a deity: a transpersonal/spiritual dimension to the mystery of life in nature – a Mother Goddess. Jungian analysts Baring and Cashford argue these figures represent the experience and mystery of our planet giving birth to life (Baring and Cashford, 1991, pp. 1–45).

What happened to this awed reverence? How did nature become a threat to be tamed, crushed, and destroyed? These authors point out that at the time these figures of the Earth Mother were made, hunting scenes were also being painted. Our ancestors balanced their need to treat the mystery of life with reverence with the need to kill to survive. At some point, the balance broke down, which I will explore later. Other species, especially predatory species, have not suffered this way. The difference between them and us appears to be our development of technology.

Looking back so many thousands of years later at these earliest figures, it seems as if humanity's first image of life was the mother. This must go back to a time when human beings experienced themselves as the children of nature, in relationship with all things, part of the whole (ibid., p. 9).

A collective attachment disorder and primary defences

Given our original attachment to the Mother Goddess, our relationship to her in the ecological crisis can be understood as an attachment disorder. Development of a healthy attachment to mother, or someone in a mothering role, is essential during the first few months of life. A 'safe base', a healthy attachment, is a retreat when under threat and feeling vulnerable. This was first recognised by American psychoanalyst Mary Ainsworth (1967). While this is complex at a collective level, an anxious avoidant neurotic attachment is the closest fit for our relationship to the Mother Goddess. In children, this arises when the person in the mothering role, insensitive to the child's needs, rejects demands and takes over in an intrusive manner (Ainsworth et al., 1978). It is difficult to see how we could pin responsibility for our neurotic attachment on the Mother Goddess herself. At an adult rational level, we made collective decisions which harmed our relationship with nature and established a rift; at a regressed level, it feels as if we were abandoned by nature because of this rift. Collective responsibility for this abandonment has been projected onto the natural world – the Mother Goddess is being a bad Mother, a hateful Goddess. In individual therapy this is like patients who decide to take a holiday then experience the therapist as abandoning them because – from a regressed infantile position – that's what it feels like. A collective sense of betrayal is in part responsible for our destructiveness.

We can understand the destructive behaviour of Western consciousness as toxic attachments derived from primary defences including 'infantile oral, urethral and anal-sadistic phantasies and impulses to attack . . . including the projection of excrements' (Klein, 1988, pp. 143, 155). Greed and envy are apparent in ripping, tearing the breast of the Mother Goddess: open cast mining, destruction of great swathes of the Indonesian rainforest for the profit to be made from palm oil plantations, destruction of the Amazonian rain forest for cattle ranching, consistent overfishing. Anal and urethral sadism, the projection of excrements to despoil, can be seen in oil spills, pollution of rivers with sewage so they become devoid of life, poisoning through widespread pesticide use (Carson, 1962), dumping of toxic waste and radio-active material, resulting in a catalogue of destruction to both the ecological systems and species diversity, treating the natural world with contempt – effectively, pissing on it.

The work of conservationists and environmentalists producing a series such as Blue Planet 2, narrated by the famous BBC wildlife presenter Sir David Attenborough, has brought the beauty of the natural world and the damage we are doing to collective awareness. Movies such as *Avatar* (Cameron, 2009) and *The Emerald Forest* (Boorman, 1985) show a heartless, murderous destructiveness toward nature; *Gorillas in the Mist* (Aptel, 1988) and *Deep Water Horizon* (Berg, 2016) reflect the tragic cost of greed. But the fact that films like these are made and become popular is a cause for optimism. Overcoming defensive denial and suppression facilitates appreciation and gratitude for the natural world, sadness at the damage we have inflicted, and an urge for reparation which is essential for the healing required.

The impact of agriculture

> All our environmental problems become easier to solve with fewer people, and harder – and ultimately impossible – to solve with ever more people.
>
> Sir David Attenborough – BBC wildlife broadcaster

Attunement to the ecological system to which a species belongs is seen throughout nature. It may be an adaptation to limit population size to the availability of food, but it seems more an understanding of their place in the web of life. We have lost touch with this.

Dr Yuval Harari, a professor of history from the University of Jerusalem who specialises in world history, suggests that our early ancestors subsisted as hunter-gatherers for 2.5 million years. Population remained controlled because of hormonal and genetic mechanisms designed to limit procreation. If food was plentiful, females reached puberty earlier; when food was in short supply, puberty was late and fertility decreased. Babies and small children who move slowly were a burden on nomadic foragers, so people tried to space their children several years apart. Women suckled their children continuously until the children were 3 or 4 years old, thereby reducing the chances of pregnancy. Cultural taboos regarding pregnancy, abortions, and infanticide also had a part to play. With the gradual introduction of agriculture ten thousand years ago, all this changed.

Ultimately, the time and effort involved in tending crops meant people gave up their nomadic lifestyle, enabling women to have a child every year. Babies were weaned earlier, so they could lend a hand with the farming, but were also another mouth to feed (Harari, 2011, pp. 87, 94, 95). The 'Luxury Trap' he describes involved the gradual erosion of the quality of life to less than they had as nomadic hunter-gatherers. Hard labour with the promise of a bumper harvest, undermined health, increased infant mortality, the diet was less diverse and so health suffered, crops invited pests such as rats, and diseases (ibid., pp. 96–8). Western society remains locked in a version of the same 'Luxury Trap', whereby the promise of better times as a reward for effort continues, deferred gratification. As Harari points out, 'Once people get used to a certain luxury, they take it for granted. Then they begin to count on it. Finally, they reach a point where they can't live without it' (ibid., p. 98).

Splits and suppression

Iain McGilchrist, a former consulting psychiatrist and clinical director at the Bethlehem Royal and Maudsley Hospital, charted a progressive shift from right brain to left brain dominance over the last two thousand years (McGilchrist, 2009, pp. 240–427). As he explains,

> the essential difference between the right hemisphere and the left hemisphere is that the right hemisphere pays attention to the Other, whatever it is that exists apart from ourselves, with which it sees itself in profound relation. It is deeply attracted to and given life by, the relationship, the betweenness, that exists with this Other. By contrast, the left hemisphere pays attention to the virtual world that it has created, which is self-consistent, but self-contained, ultimately disconnected from the Other, making it powerful, but ultimately only able to operate on, and know, itself.
>
> (ibid., p. 93)

In *Memories, Dreams, Reflections*, Jung (1983) refers to a split within himself which he first recognised as a child and was later to identify as his personality No. 1, which was positivistic, and his personality No. 2, which was romantic. This Otherness within himself, and his wish to understand it, fuelled a passion which was to become his life-long quest. Papadopoulos argues the 'Jungian opus could be appreciated more fully: 'as a series of reformulations of his understanding of the Other' (Papadopoulos, 2002, p. 13). As a boy, Jung was in touch with a pantheistic sense of the divine in nature through imaginal perception facilitated by his No. 2 personality – which he later explored in the context of *lumen natra* – the light of nature. However, Jung tells us this experience had no cultural articulation at the time and, by his mid-teens, it had become 'a remote and unreal dream' (Jung, 1983, p. 68).

Roger Brooke, Professor of Psychology at Duquesne University, argues for Jung as for us, the Western educational process 'recapitulates, in certain fundamental ways, the historical drama of western consciousness, particularly as it unfolded in the renaissance' (2000, p. 13). Robert Romanyshyn, Emeritus Professor of Psychology at Pacifica Graduate Institute California, argues:

> The invention of linear perspective in painting in fifteenth century Italy has become "the cultural vision which has shaped our contemporary technological world . . . the viewer is imagined to be looking at the world as if through a window. This window has become our habit of mind and through it we have become a self which has learned to keep its eye upon the world. Behind the window we have become distant and detached, a self-separated and isolated from the world.
>
> (Romanyshyn, 1989, pp. 32, 67)

The split in Jung's psyche needs to be seen in the wider context of the split in Western psyche. The emergence of Romanticism at the end of the nineteenth century was evident through art, literature, the intertwining of psychology with spiritualism, and the 'passionate popular interest in . . . spiritualism, which had spawned both parlour games and conscientious investigating bodies' (Haule, 1984, 239). The need for this romantic resurgence in the collective psyche of Western civilisation can be seen partly as a compensation for the 'controlled rationalism of the eighteenth century' (Singer, 1973, pp. 8–9).

We can see the struggle Jung evidently had in accommodating the different influences of his fractured psyche, also apparent in his *Red Book* in which he refers to the 'spirit of the time' and the 'spirit of the depths' (Jung, 2009). 'The division of the world into objective and subjective categories (scientific and personal, rational and irrational) is also a self-division . . . The Renaissance division of the world was thus deeply felt by Jung as a fracture within himself' (Brooke, 2000, p. 14). This fracture reflects the split from the imaginal perception and understanding of the natural world, once felt as an inseparable intimate reality.

The term imaginal comes from a paper published in 1972 by the Islamic scholar Henry Corbin in a paper entitled 'Mundus Imaginalis, or the Imaginary and the Imaginal' (1972). The imaginal is essentially non-Egoic imagination, and so distinct from imaginary or Ego-generated fantasy (See chapter 8 in this volume). Free of the Ego's control and censorship, it provides perception via alternative states of consciousness, and its nature is metaphor rather than literal. It is found by the lowering of Ego into the unconscious, resulting in an altered state of consciousness (Field, 1992), a waking dream (Watkins, 1976), and as such involves a process of dissociation.

Technological man as creator god

Jung was clearly sensitive to the role of nature in providing the nourishment a human soul needs and its role in bridging to the divine. 'Nature, psyche, and life appear to me like divinity unfolded – what more could I ask for?' (Sabini, 2008, p. 221), but the following quote describing an experience while travelling through Kenya and Uganda presents a perspective which is indicative of an inflated split from nature.

> Man, I, in an indivisible act of creation puts the stamp of perfection on the world by giving it objective existence . . . man is indispensable for the creation of creation; . . . he himself is the second creator of the world, who alone has given the world its objective existence – without which, unheard, unseen, silently eating, giving birth, dying, heads nodding through hundreds of millions of years, it would have gone on in the profoundest night of non-being down to its unknown end. Human consciousness created objective existence and meaning, and man found his indispensable place in the great process of being.
>
> (Jung, 1983, pp. 284–5)

This is hugely arrogant. We have been around for hardly the blink of an eye on the scale of evolutionary development. The natural world is constantly exploring and expressing itself though developing new species with their themes and variations. We are simply one of the more recent experiments. Nature does not need us to exist, and the idea human consciousness recreates creation is a delusion. This madness is not Jung's alone. It is the madness of the collective consensus of Western consciousness, powerfully expressed in Mary Shelley's novel 'Frankenstein' (Shelley, 1818/2003) (and see chapter 6 of this volume). Her novel, a product of Romanticism, is about denial of mortality, and a man claiming the power to cheat death by creating a monster. This is the same as Jung's argument about humans (which he refers to as 'Man') recreating creation by superior consciousness – humans become God, mortality is denied, and humans claim the power to cheat death through the creation of a monster. The parallel with the development of technology and then denying dependence on the nature is easy to see. Technology is not bad or wrong of itself. It is the product of a creative dream from psyche, but 'in living these dreams forgetfully, in tending only to the surface of technology as event while forgetting its imaginal lining, these dreams can become a nightmare of destruction' (Romanyshyn, 1989, p. 13).

Disembeddedness and the Ego–Self axis

We have seen there is a progressive distancing from an intimate relationship with nature at our earliest beginnings, particularly in the shift from hunter-gather subsistence through the distancing from a grounded relationship with nature and the adoption of linear perspective in the Renaissance. This ushered in the arrogance of the Enlightenment and the suppression of the imaginal. In today's world, maintaining a relationship with nature is still harder because of the rate of change in our lives (see Chapter 11 of this volume) and the undermining of our fundamental sense of identity and security.

The consumer culture of the globalised Western world is no longer bound by geography. Mass media has built a 'global village' where the Western mind set is increasingly intruding into lands where it previously had no influence. Jock Young, Professor of Sociology at the University of Kent, argues:

> mass migration, tourism, the flexibility of labour, the breakdown of commu-
> nity, the instability of family, the rise of virtual realities and reference points
> within the media as part of the process of globalisation, the impact of mass
> consumerism, and the idealisation of individualism, choice and spontane-
> ity . . . have been ratchetted up in the present period.
>
> (Young, 2007, pp. 2–3)

This has generated disembeddedness: a crisis of ontological insecurity and pre-cariousness of being (ibid.), as identity is itself treated as a commodity to be pur-chased in the context of a consumer culture. American psychiatrist and Jungian analyst Edward Edinger described the development of an Ego–Self axis during infancy:

> the ego-self axis represents the vital connection between ego and self that
> must be relatively intact if the ego is to survive stress and grow. This axis
> is the gateway or path of communication between the conscious personal-
> ity and the objective psyche. Damage to the ego-self axis impairs or destroys
> the connection between conscious and unconscious, leading to alienation of
> the ego from its origin and foundation.
>
> (1960, p. 8)

The Self is experienced as an empirical deity and subordinates the Ego to it. Our first experience of this deity is through projection onto mother: mother = God, the all-powerful centre of our universe. This axis becomes conscious during indi-viduation, by which consciousness is developed through a dialectic relationship between Ego and Self, and 'is vitally important to psychic health. It gives founda-tion, structure and security to the ego and provides energy, interest, meaning, and purpose' (p. 43). If the early relationship with the primary caregiver is faulty in some way, through illness, depression, or abuse, for example, the necessary expe-rience of acceptance and reparation is not received, leaving the infant to feel cast out, empty, wrong, bad, alienated. This leads to feelings of despair, and murder-ousness. These murderous feelings can be turned inwards or acted out in relations with others (Edinger, 1972, p. 43).

We can understand the Ego–Self axis as operating in the relationship between our collective consciousness and the Self as originally experienced in our relation-ship with our Mother God, then transformed to 'mother nature'. Frustrations in this relationship result from development of technology, leaving us feeling pro-gressively abandoned. The resultant murderous rage has intensified over recent decades. I suggest that the disembeddedness identified by Young is essentially alienation experienced at a social level.

Parallels with eating disorders – consuming the planet

Eating disorders form a spectrum. At one pole is anorexia which seeks denial of need, independence, and an escape from the body which is attacked as if it is mother. Dependent needs and feelings are a threat. Through denial of the body, its mortality as part of the natural world is denied. At the other extreme compulsive eating involves excessive consumption to try to fill a deep emotional emptiness. Bulimia holds the ambivalent middle ground. Eating disorders reflect our relationship with nature, acted out by the sufferers in their own bodies: 'the mandate for discipline clashes with the mandate for pleasure, and our bodies serve as the "ultimate metaphor" reflecting the general mood and cultural contradictions of late capitalist society' (Crawford, 1998). Whilst bulimia expresses the unstable double-bind of consumer capitalism, anorexia (i.e., work ethic in absolute control) and obesity (i.e., consumerism in control) embody an attempted 'resolution' of these cultural contradictions (Williams and Bendelow, 1998, p. 75).

Yet bingeing, and particularly compulsive eating, at the extreme is not be pleasurable. Consumption is pushed beyond the body's limit of containment, resulting in a pain and increasing distress, which serves an unconscious persecutory agenda. But whereas the body of the individual is forced to suffer, the natural world suffers the persecutory excesses of Western culture.

Compulsive consumption happens increasingly in our culture through consumerism and 'retail therapy'. It is a manic avoidance of our collective starvation and emptiness, a rupture in our relationship with the natural world. Consumerism is driven by a fear of emptiness and denial of inner psychological reality. It provides an illusion of control over ravenous hunger, but the image of the anorexic mirrors the true nature of the shadow of Western consciousness; starvation and domination of nature within the body, fighting to deny the natural need for love and nourishment from the Mother Goddess. The metaphor of our unconscious plight is made literal in the body of a person suffering from severe anorexia:

> anorexia is a symptom by which the psyche is trying to alert us to the consequences of contempt for depth connection with our own nature . . . the dysfunctional family dynamic of an anorexic acts as a conduit for societal shadow, and the anorexic individuals are effectively elected to mirror our societal malaise.
>
> (Colverson, 2014, p. 246)

When the symbolic nature of its message is not heeded, psyche has a habit of making it literal. The image anorexia presents is a prediction of the starvation we could all suffer as a result of over-population – when as a result of the ever-increasing demand the breasts of the Mother Goddess finally run dry.

Shamanic healing

> The psyche is not of today; its ancestry goes back many millions of years. Individual consciousness is only the flower and fruit of a season. Sprung from the perennial rhizome beneath the earth; and it would find itself in better accord with the truth if it took the existence of the rhizome into its calculations. For the root matter is the mother of all things.
>
> (Jung, CW 5, Foreword, p. xxiv)

> Our normal waking consciousness, rational consciousness as we call it, is but one special type of consciousness, whilst all about it, parted from it by the filmiest of screens, there lie potential forms of consciousness entirely different.
>
> (William James, 1902/1985, p. 283)

Shamanism offers a mode of healing Western consciousness has largely forgotten, and which is particularly pertinent to the 'lost soul' of Western consciousness. It is the oldest form of healing we know of and began over thirty thousand years ago among the Tungus peoples of north East Asia and Siberia. The root of saman, 'sa', is 'to know': a shaman is one who knows (Jakobson, 1999, p. 3), and the term shaman refers to 'persons of both sexes who have mastered spirits, who at will can introduce these spirits into themselves and use their power over the spirits in their own interests, particularly helping people' (Shirokogoroff, 1935/1982, p. 269). He/she 'becomes both a connective tissue to the invisible realm of spirits and a technician of the supernatural' (Thorpe, 1993, p. 29).

Among the finds of cave paintings across Europe mentioned earlier were those found in the Chauvet Cave on the Ardeche River, a tributary of the Rhone in Southern France. Some images here date before 30,000 B.P. – the date of the Mother Goddess figures noted earlier.

> One of the most impressive figures in this cave is a composite figure, now known as the "Sorcerer", with a head and upper body of a bison and the lower extremities of a human – a figure which very much resembles certain images of the shaman as he has survived into historical times.
>
> (Ryan, 2002, p. 194)

The role of shaman may have been passed on from a parent, or they may have been 'called' to it through a traumatic event or illness. Either way training involves a period of trial, and experience of suffering. Groesbeck argues the shaman can be regarded as a wounded healer, *par excellence*, and this archetype has its origin in shamanic experience (Groesbeck, 1988).

Shamanic healing may involve gathering and preparation of herbs for medicinal use, exorcistic and blessing rites and songs, prayers, dances, and incantations, 'but the heart of the healing effort . . . is the shamanic trance, a journey into another reality outside time and space' (Sandner, 1997, p. 3). This trance is auto-hypnotic and may vary in depth as required. It is characterised: 'by dissolving of the boundaries of the mundane world in which every day things are but fragments of existence artificially removed from the wholeness of being and then given a relative sense of meaning; entities that do not really exist for us except in a conditioned way' (Kalweit, 1992, p. 72).

The most important object for a shaman (in most cultures) is his/her drum, its rhythmic beat facilitating dissociation and a shift into an altered state of consciousness. It symbolises the universe as well as countless other things, and may enable the shaman 'to fly through the air, summon and "imprison" the spirits, or enable the shaman to concentrate and regain contact with the spiritual world through which [he/she] is preparing to travel' (Eliade, 2004, p. 168). Its iconography is dominated by the symbolism of the ecstatic journey which implies 'a break-through in plane' (ibid., p. 173).

Shamans provide a service to the community by healing sickness, rescuing lost souls, and caring for everyone's spiritual welfare by journeying within the three worlds of the shamanic cosmos: sky (upper), earth (middle), and underworld (lower).

> The upper and lower worlds are transcendent realities, so-called non-ordinary realities, which we are not trained to see when we are in our normal waking state of consciousness but can be accessed in non-ordinary states. The middle world is twofold: it consists of the energetic aspects (dream aspects) of the everyday world we live in, and its visible, physical aspects in the form of the worlds we experience daily with our five senses. . . . All four worlds are of equal importance, forming together the description of the whole.
>
> (Mackinnon, 2012, p. 191)

The non-ordinary states which Mackinnon refers to involve imaginal perception, through which tutelary spirits are employed as guides on the journey. Some spirits take animal form: bears, wolves, stags, hares, birds of all kinds; some are divine or semi-divine beings (Eliade, 2004, pp. 88, 89). The reality of these other worlds is taken as being as real as waking consciousness by the shaman.

An axis travels through an 'opening', a 'hole' between the worlds; through this the gods descend to earth and the dead go to the subterranean regions. Through the same hole, the soul of the shaman in ecstasy can fly up or down during their journeying (Eliade, 2004, p. 259). This axis, known as the World Tree in shamanic cosmology, is central and essential to the ideology and experience. 'From its wood he makes his drum; climbing the ritual birch, he effectively reaches the summit of the Cosmic Tree; in front of his yurt and inside are replicas of the Tree and he depicts them on his drum' (ibid., p. 270).

Cosmologically, the World Tree rises at the centre of the earth, the place of earth's umbilicus and the branches touch the heavens. The World Tree represents the universe in continual regeneration, the inexhaustible spring of cosmic life, the paramount reservoir of the sacred; and it symbolises the sky and the planetary heavens. It is related to ideas of creation, fecundity, and initiation, and finally to the idea of absolute reality and immortality (Eliade, 1964, pp. 270–1).

The shaman penetrates the root matter of all things which lies beyond the plane of consciousness, 'which our senses can grasp only the outer sheath or veil and does so with therapeutic and revitalising effect' (Ryan, 2002, p. 35). A Sioux shaman, for example, refers to himself as 'interpreter' and medium for the primordial force, in his interpretations and visions it is in fact the world force which is speaking (Powers, 1984, p. 56). This example from the Sioux nation is typical; shamanism bears 'strikingly precise structural and functional similarities in areas of the world where possible cultural exchange appears extremely unlikely or extraordinarily remote in time' (Ryan, 2002, p. 37), suggesting an archetypal core.

Eliade points out that

> we get the impression that for archaic societies life cannot be repaired, it can only be re-created by a return to the sources. The "source of sources" is the prodigious outpouring of energy, life, and fecundity that occurred at the creation of the world.
>
> (Eliade, 1964, p. 30)

There is a parallel here between the Ego–Self axis identified by Edinger as being between Ego and Self, and the axis between the worlds employed by shamans. While presented in different contexts, both provide access to transpersonal realms. We can understand the issues of disembeddedness and alienation described earlier as the damage to our ability to access the connectedness and oneness experienced by shamans in their journeys via the World Tree, a perception which was once very much part of our experience.

Non-dual reality

> There came one and knocked at the door of the Beloved.
> And a voice answered and said, 'Who is there?'
> The lover replied, 'It is I.'
> 'Go hence,' returned the voice;
> 'there is no room within for thee and me.'
> After a year of solitude and deprivation this man returned to the door of the
>   Beloved and knocked and again the voice demanded,
> 'Who is there?'
> He answered, 'It is thou'.
> 'Enter,' said the voice, 'for I am within'.
>
> Rumi (thirteenth-century Sufi poet and mystic)

Jung collaborated with the Nobel prize-winning physicist Wolfgang Pauli on development of his ideas on synchronicity: simultaneous occurrence of events which appear significantly related but have no discernible causal connection (see Chapter 11 of this volume). This expanded Jung's ideas regarding psyche. He recognised that 'it is not only possible but highly probable, even, that psyche and matter are two different aspects of one and the same thing' (Jung, CW 8, para. 418). From this, Jung expanded his understanding of archetypes and described their 'psychoid nature' using an analogy with the electromagnetic spectrum, where the high frequency ultra-violet end described a spiritual dimension, and the low frequency infra-red described psyche merging into physical matter (ibid., para. 420). This understanding has considerable ramifications.

I think the palaeontologist and French Jesuit Teilhard de Chardin (1959/1966) put it more succinctly: 'Matter is spirit moving slowly enough to be seen' (p. 53ff). By logical extension then, everything in the entire cosmos is psyche, in different energetic forms of manifestation, which is maybe why Jung, towards the end of his life, began to coin the term objective psyche – psyche is not a personal possession, it is something we are all in and all part of.

In Jung's map of the psyche showing the various layers of the unconscious down through personal Ego, family, community, nation. . . , the ultimate depth he calls 'the central force' (Jacobi, 1973, p. 34). I think this central force is universal consciousness, which is ultimately what Jung meant by the Self, though it took most of his life for his thinking to catch up with his initial intuition. The Self as Jung defined it is the centre and circumference of the psyche – the entire cosmos – and everything within it. It is experienced as numinous because it is divine, though Western consciousness does not perceive this.

The 'creation time' of shamanism can be understood as the non-dual nature of universal consciousness. This has been rigorously investigated by Dean Radin, senior scientist at the Institute of Noetic Sciences, in the context of paranormal research (Radin, 2006, 1997/2009). The non-dual is found in ancient spiritual philosophies in Adivaita Vedanta, Taoism, Sufism, Buddhism, and some Christian mystics (Loy, 1997). While existing as individuals with separate bodies and Egos at one level, at another we are all participate in an underlying level of transpersonal consciousness, where we are not separate (Corbett, 2011, p. 236).

The popular positivistic understanding of consciousness is of something generated by the brain. An alternative understanding is 'consciousness is a superordinate, irreducible principle, the common ocean in which we swim, the larger mind behind the personal mind . . . so at the deepest level we all participate in the same consciousness' (ibid., p. 248), a notion supported by the evidence such as ESP, and near death experiences in which consciousness continues after recorded brain death (see particularly Kelly and Kelly, 2010, pp. 392, 393). As Aldous Huxley put it:

> To make biological survival possible, Mind at Large has to be funnelled through the reducing valve of the brain and the nervous system. What comes

out the other end is a measly trickle of the kind of consciousness which will help us to stay alive on the surface of this particular planet.

(Huxley, 1954/1990, p. 22)

At a non-dual level, we are depleted as we extinguish more and more species; we are hurt, damaged, and endangered, because we are the natural world which we attack and despoil. Like a dissociated borderline person who cuts herself, we need to recognise the terrific damage we have done to the natural world is something we have done to ourselves and speaks of an inner pain we need to address.

For indigenous cultures such as native Australians, who are more in tune with the non-dual nature of reality, 'landscape is at the centre of everything: at once the source of life, the origin of the tribe, the metamorphosed body of blood-line ancestors and the intelligent force which drives the individual and creates society' (King-Boyes, 1977), in this context landscape is the 'earth-sky-water-tree-human continuum' which is the cosmological and existential ground of the Aboriginal Dreaming (Wright, 1985, p. 32). 'Nature is never "only natural"; it is fraught with religious value . . . the world is impregnated with sacredness . . . (with) different modalities of the sacred in the very structure of the world and cosmic phenomena' (Eliade, 1959, p. 116).

In Australia, following colonisation, the doctrine of 'terra nullius' – a land without people – was established under British colonial government and persisted in Australian law until 1992. It served to reinforce the concept that indigenous land was 'empty'; it belonged to no one and so could rightly be claimed for Western exploitation or settlement (Shannon, 2016). This can be understood as a highly destructive envious attack by these Western powers. Indigenous people were forced from their traditional land with which they had an intense spiritual connection, a traumatic experience sending many into deep depression, and drug addiction – predominantly alcohol, a concretised relationship with spirit. It is tragic that such a culture, which had been invested in a 'symbolically saturated sacred conscious reality' for millennia (Tacey, 2009, p. 188), should be ripped out of it in this way. Suicide was virtually unknown among indigenous Australians prior to the 1980s, but it has now reached crisis levels (Koff, 2019).

As Eliade points out:

The man of traditional societies . . . can live only in a sacred world, because it is only in such a world that he participates in being, that he has *real existence*. For him profane space represents absolute non-being. If, by some evil chance, he strays into it, he feels emptied of his ontic substance, as if he were dissolving into chaos, and he finally dies.

(1959, p. 64; italics in original)

This is the condition of Western consciousness. It has gone largely unnoticed because it happened over a protracted period. We adopt behavioural strategies, such as consumerism and addictions, to distract us from this reality (ibid.), but from a shamanic perspective, Western consciousness is suffering from a loss of soul.

Conclusion

> Prayer is not an old woman's idle amusement. Properly understood and applied, it is the most potent instrument of action.
>
> (Mahatma K. Gandhi, 1946, p. 195)

We have no simple solution to an issue which has grown in strength over millennia. It is not a problem which Western consciousness can solve because Western consciousness is the issue. We cannot turn back the clock and become as hunter-gatherers – though this may happen by default if the worst predictions come true. The work of conservationists, activists, ecologists, and wildlife photographers, writers, and presenters, is imperative to keep the wonder of and concern for the natural world clearly in Western consciousness.

Drawing on a holographic model of the psyche (Zinkin, 1987, pp. 1–21), a notion which fits well with the non-dual, we all have an inner shamanic healer within us. Our collective memory of shamanic healing creates a strong morphic field (Sheldrake, 1981). In response to the current crisis, it is possible for a new morphic field can grow from the roots of this old one (Sheldrake, 2019, personal communication), which may facilitate the necessary shift in Western consciousness

The Zen Buddhist sage and author D.T. Suzuki argues the burnished mirror of non-dual reality stands before us; we only need to investigate it (Suzuki, 1998). A glimpse by enough people may bring a waking dream of a different narrative into Western consciousness, through which we can recognise we are one with the natural world.

There is cause for optimism. Eliade describes a Ying-Yang relationship between the sacred and profane (Eliade, 1959). Desacralising attitudes have (hopefully) reached their zenith, and attitudes towards the natural world are changing. The emerging Borderland consciousness to which Bernstein refers (2005) is becoming more evident, particularly through recent climate change protests. We do not need a swing to the other extreme and demonise technology. We need a balanced relationship whereby technology is practised and developed within a fully conscious relationship to the sacred nature of the natural world, of which we are a part. So, like an eddy in a river, Western consciousness can maintain its identity as a unique phenomenon, while also being part of the magnificent whole. Through tuning into an imaginal perception, we can recognise that there is no Other.

References

Ainsworth, M.D.S. (1967). *Infancy in Uganda: Infant care and the growth of attachment.* Baltimore, MD: John Hopkins University Press.

Ainsworth, M.D.S., Blehar, M., Waters, E. and Wall, S. (1978). *Patterns of attachment: A psychological study of the strange situation.* Hillsdale, NJ: Lawrence Erlbaum Associates.

Aptel, M. (1988). *Gorillas in the mist*. Los Angeles: Universal Pictures.

Barber, E.A. (1968). *Liddel, Scott and Jones, Greek – English Lexicon: A supplement*. Oxford: Clarendon Press.

Baring, A. and Cashford, J. (1991). *The myth of the goddess: Evolution of an image*. St. Ives: BCA.

Berg, P. (2016). *Deep water horizon*. Santa Monica: Summit Entertainment, Lionsgate Films.

Berstein, J. (2005). *Living in the borderland: The evolution of consciousness and the challenge of healing trauma*. London and New York: Routledge.

Boorman, J. (1985). *The emerald forest*. Los Angeles: Christel Films, Embassy Pictures.

Brooke, R. (2000). Jung's recollection of the life-world. In: *Pathways into the Jungian world: Phenomenology and analytical psychology*. London: Routledge.

Cameron, J. (2009). *Avatar*. Los Angeles: 20th Century Fox.

Carson, R. (1962). *Silent spring*. New York, NY: Houghton Mifflin Harcourt Publishing Company.

Colverson, J. (2014). Anorexia and alchemy. In: D. Mathers, ed., *Alchemy and psychotherapy: Post Jungian perspectives*. London: Routledge.

Corbett, L. (2011). *The sacred cauldron: Psychotherapy as a spiritual practice*. Wilmette, IL: Chiron Publications.

Corbin, H. (1972). *Mundus imaginalis, or the imaginary and the imaginal*. New York: Spring.

Crawford, R. (1998). The ritual of health promotion. In: S.J. Williams, J. Gabe, and M. Calnan, eds., *Theorising health, medicine and society*. London: Sage.

Edinger, E.F. (1960). The ego-self paradox. *Journal of Analytical Psychology*, 5(1), pp. 3–18.

———. (1972). *Ego and archetype*. Boston, MA: Shambhala Publications, Inc.

Eliade, M. (1959). *The sacred and the profane*. New York, NY: Harcourt, Bruce & World.

———. (1964). *Myth and reality*. New York, NY: Harper & Row Inc.

———. (2004). *Shamanism: Archaic techniques of ecstasy*. Princeton, NJ: Princeton University Press.

Field, N. (1992). The therapeutic function of altered states. *Journal of Analytical Psychology*, 37, pp. 211–234.

Gandhi, M. (1946). *Collected works of Mahatma Ghandi*, Vol. 90, doc. 253. New Delhi: Publications Division of the Ministry of Information and Broadcasting of the Govt. of India.

Groesbeck, C.J. (1988). *The archetype of the healer*. Unpublished Manuscript.

Harari, Y.N. (2011). *Sapiens: A brief history of mankind*. London: Vintage.

Haule, J.R. (1984). From somnambulism to the archetypes: The French roots of Jung's split with Freud. *Psychoanalytic Review*, 71(4).

Huxley, A. (1954/1990). The doors of perception. In: *The doors of perception and heaven and hell*. New York, NY: Perennial Library and Harper & Row, pp. 7–79.

Jacobi, J. (1973). *The psychology of C.G. Jung*. Newhaven, CT: Yale University Press.

Jakobson, M.D. (1999). *Shamanism: Traditional and contemporary approaches to the mastery of spirits and healing*. New York, NY and Oxford: Berghahn Books.

James, W. (1902/1985). *Varieties of religious experience*. Cambridge, MA: Harvard University Press.

Jung, C.G. (1953–77). *Except where indicated, references are by volume and paragraph number to the collected works of C.G. Jung*. 20 vol., ed. by H. Read, M. Fordham, and

G. Adler, trans. by R.F.C. Hull. London: Routledge and Princeton: Princeton University Press.

———. (1983). *Memories, dreams, reflections*. London: Fontana Paperbacks.

———. (2009). *The red book: Liber Novus*, ed. by S. Shamdasani. New York, NY and London: Norton and Company.

Kalweit, H. (1992). *Dreamtime and inner space: The world of the shaman*. Boston, MA: Shambala Publications.

Kelly, E.F. and Kelly, E.W. (2010). *Irreducible mind*. Lanham, MD: Rowman and Little-field Publishers Inc.

King-Boyes, M. (1977). Creation and cosmology: Man and nature. In: *Patterns of aboriginal culture: Then and now*. Sydney: McGraw Hill.

Klein, M. (1988). *Envy and gratitude and other works 1946–1963*. London: Virago.

Koff, J. (2019). Available at: www.creativespirits.info/aboriginalculture/people/aboriginal-suicide-rates [Accessed 4 Aug. 2019].

Loy, D. (1997). *Nonduality: A study in comparative philosophy*. Amherst, NY: Humanity Books.

Mackinnon, C. (2012). *Shamanism and spirituality in therapeutic practice: An introduction*. London and Philadelphia, PA: Singing Dragon.

McGilchrist, I. (2009). *The master and his emissary: The divided brain and the western world*. New Haven, CT and London: Yale University Press.

Papadopoulos, R.K. (1984). Jung and the concept of the other. In: R.K. Papadopoulos and G. Saayman, eds., *Jung in modern perspective*. London: Wildhouse.

———. (2002). The other other: When the exotic other subjugates the familiar other. *Journal of Analytical Psychology*, 47(2), pp. 163–188.

Powers, W.K. (1984). *Oglala religion*. Lincoln, NE: Bison Books and University of Nebraska Press.

Radin, D. (1997/2009). *The conscious universe: The scientific truth of psychic phenomena*. New York, NY: Harper Collins.

———. (2006). *Entangled minds: Extrasensory experiences in a quantum reality*. London, New York and Sydney: Paraview Pocket Books.

Romanyshyn, R. (1989). *Technology as symptom and dream*. London and New York: Routledge.

Ryan, R.E. (2002). *Shamanism and the psychology of C.G. Jung: The great circle*. London: Vega.

Sabini, M., ed. (2008). *The earth has soul: C.G. Jung on nature, technology & modern life*. Berkeley, CA: North Atlantic Books.

Sandner, D.F. (1997). Analytical psychology and shamanism. In: D.F. Sandner and S.H. Wong, eds., *The sacred heritage: The influence of shamanism on analytical psychology*. London and New York: Routledge.

Shannon, D.S. (2016). Available at: http://theconversation.com/refugees-in-their-own-land-how-indigenous-people-are-still-homeless-in-modern-australia-55183 [Accessed 21 July 2019].

Sheldrake, A.R. (1981). *A new science of life: The hypothesis of formative causation*. London: Blond and Briggs.

Shelley, M. (1818/2003). *Frankenstein: or the modern prometheus*. London: Penguin Classics.

Shirokogoroff, S.M. (1935/1982). *Psychomental complex of the Tungus*. London: Kegan Paul, Trench, Trubner & Co., Ltd.

Singer, J. (1973). *The child's world of make believe: Experimental studies of imaginative play*, ed. by L. Jerome, Child Psychology Series. New York: Academic Press.

Suzuki, D.T. (1949/1998). The Zen doctrine of no-mind. In: A. Molino, ed., *The couch and the tree: Dialogues in psychoanalysis and Buddhism*. London: Open Gate Press.

Tacey, D. (2009). *Edge of the sacred: Jung, psyche, earth*. Einsiedeln, Switzerland: Daimon Verlag.

Teilhard de Chardin, P. (1959/1966). *Issues in science and religion*, ed. by I.G. Barbour. Englewood Cliffs, NJ: Prentice Hall, p. 53ff.

Thorpe, S.A. (1993). *Shamans, medicine men and traditional healers: A comparative study of shamanism in Siberian Asia, southern Africa and North America*. Pretoria: Sigma Press.

Watkins, M. (1976). *Waking dreams*. New York: Interface Publications.

Williams, S.J. and Bendelow, G. (1998). *The lived body: Sociological themes, embodied issues*. London and New York: Routledge.

Wright, J. (1985). Landscape and dreaming. In: Stephen R. Graubarb (ed.), *Australia: The Daedalus Symposium*. Sydney: Angus & Robertson.

Young, J. (2007). *The vertigo of late modernity*. London, California, New Delhi and Singapore: Sage Publications Ltd.

Zinkin, L. (1987). The hologram as a model for analytical psychology. *Journal of Analytical Psychology*, 32, part 1, pp. 1–21.

Chapter 10

Our connection with animals and the universe

Psychology, symbols, spirituality and the new physics

Ruth Williams

We all love animals. We feel connected to them in deep and mysterious ways which are not always conscious. Those of us lucky enough to live with domestic animals are privileged to feel the devoted companionship they offer, and often witness their uncanny wisdom. Parents will often lull their children off to sleep with animal fables. They take you into the realm of the imaginal. Even Machiavelli used animals as a metaphor when he said a successful Prince had to be like both the lion and the fox (Samuels, 1993, p. 79). People spend hours glued to online videos of cats and dogs as a way of feeling a connection to this realm of soul which can feel so absent in the digital age.

I aim to set out ways in which we are all connected to the more-than-human world and how loss of this link causes us to feel so disconnected both from ourselves and the world around us. Western societies – with their ever-spinning demands to act swiftly and to keep ourselves locked in to our screens (the latest form of addiction), put untold pressure on us to think we are separate from the animal kingdom and to become ever more disconnected from our souls. How can we then appreciate the souls of those – including of the animal kind – around us? Jung was aware of this connection to the numinous in nature from an early age:

> The self is a fact of nature and always appears as such in immediate experiences, in dreams and visions . . . it is the spirit in the stone, the great secret which has to be worked out, to be extracted from nature, because it is buried in nature herself.
>
> (Jung, 1936/1989, p. 977)

Philip Pullman had it right in his trilogy *His Dark Materials* – which includes *Northern Lights* (1995), *The Subtle Knife* (1997) and *The Amber Spyglass* (2000) – when he allotted individuals animal *daimons* to act as guides and protectors. It was a brilliantly intuitive device down to the way the *daimons* were malleable and could transform into different animals as needed until adulthood when they settled into a consistent, trusty 'self'. His device highlights our own animal nature. It might be interesting to think about what your own animal *daimon* might be. I have found this imaginative exploration can also be of value in the clinical setting. In this chapter, we will be looking at: animals in dreams, animal symbolism, animals as healers, shamanic animals and therapy with animals.

The different systems conceptualising the energy fields which enable us to understand our connection to each other including to animals such as: *unus mundus*, or unified world view, including Jung's concept of the Psychoid Unconscious; Implicate Order (David Bohm); Akashic Field (Erwin Laszlo); Morphic Resonance (Rupert Sheldrake); Holographic Universe (Bohm and Pribram); and Borderland Consciousness (Jerome Bernstein). I will discuss each further.

Our own animal nature

Animals in dreams

Animals play a vital role in dreams acting as powerful symbols. An animal or insect will have quite different connotations for each of us. A different breed creates further layers of more nuanced connotations. If it is your own animal in the dream, this brings it 'closer to home' (Russack, 2002). Elsewhere, I discussed nightly encounters with dream cats which were terrifying and numinous (Williams, 2017). This series of dreams recurred over some years and were the source of deep transformation. Pioneer of archetypal psychology (offshoot of Jung's analytical psychology) James Hillman (1926–2011) puts the significance of such dreams well:

> In many cultures, animals do the blessing since they are the divinities. That's why parts of animals are used in medicines and healing rites. Blessing by the animal still goes on in our civilised lives, too. Let's say you have a quick and clever side to your personality. You sometimes lie, you tend to shoplift, fire excites you, you're hard to track and hard to trap; you have such a sharp nose that people are shy of doing business with you for fear of being outfoxed. Then you dream of a fox! Now that fox isn't merely a "shadow problem", your propensity to stealth. That fox also gives an archetypal backing to your behavioural traits, placing them more deeply in the nature of things. The fox comes into your dream as a kind of teacher, a doctor animal, who knows lots more than you do about these traits of yours.
>
> (1997, p. 2)

Jungian Analyst Neil Russack (1936–2011) from San Francisco wrote the wonderful *Animal Guides: In Life, Myth and Dreams* (2002), in which he looks at how animals come into our lives in all those realms. He explores metaphorical meanings of animals, as well as their healing power and how we can become connected to animals. He divides animals according to the elements: those associated to water (i.e., those that live in or adjacent to water), to earth (elephants, horses, deer) and to fire (dragons and unicorns, the phoenix). Curiously, he omits a chapter on air which would of course include birds and insects. Animals, Russack tells us, 'link us to the mythic realm. They are messengers of the gods' (ibid., p. 10). Like Jung before him, Russack links the loss of this association to the despoliation of the planet and our alienation from the natural world.

Animal symbolism

The canon of depth psychology is full of examples of animal symbolism. Some is familiar and in the vernacular of everyday speech. If we think or dream of a cat, do we mean a domestic moggie, a tiger, a lion, a witch's cat (Williams, 2017)? Each has a different story to tell. We can use Jung's method of amplification to deepen the meanings and reach more nuanced associations. James Hillman was an exponent of this method *par excellence*. In his *Animal Presences* (2008), Hillman discusses the symbolism of the snake. He breaks it down into twelve parts:

1   The snake is renewal and rebirth, because it sheds its skin.
2   The snake represents the negative mother, because it wraps around, smothers, won't let you go, and swallows whole.
3   It is *the* animal embodiment of evil. It is sly, shifty, sinister, fork-tongued.
4   It's a feminine symbol, having a sympathetic relation to Eve.
5   The snake is phallus, because it stiffens, erects its head, and ejects fluid from its tip. Besides, it penetrates crevices.
6   It represents the material earth world and as such is a universal enemy of the spirit.
7   The snake is healer; it is a medicine . . . It was kept in the healing temples of Asclepius in Greece, and a snake dream was the god himself coming to cure.
8   It is a guardian of holy men and wise men.
9   The snake brings fertility, for it is found by wells and springs.
10 A snake is Death, because of its poison and the instant anxiety it arouses.
11 It is the inmost truth of the body, like the sympathetic and para-sympathetic nervous systems or the serpent power of the Kunda-lini yoga.
12 The snake is the symbol for the unconscious psyche, particularly the intro-verting libido, the inward-turning energy that goes back and down and in. Its seduction draws us into darkness and deeps. It is always a 'both': creative-destructive, male-female, poisonous-healing, dry-moist, spiritual-material.

(ibid., pp. 76–7; italics in original)

He sees the last, the snake as unconscious psyche, to be the most all-encompassing.

Animal healers

Probably we have all experienced inexplicable phenomena with animals. How do they know when their owners are about to arrive home? Because they do. How do they know when their owners are ill? It is well known that dogs can detect when their owners are sick and have saved lives by alerting the owner to seek medical attention. For example, the animal is drawn to a changed smell in an area where cancer has begun. Or when they detect an owner is about to have an epileptic fit. It is worth quoting biologist Rupert Sheldrake who cites an impressive anecdote:

> [the dog] can sense, up to 50 minutes before, that I am going to have an attack and taps me twice with his paw, giving me time to get somewhere safe. He can also press a button on my phone and bark when it is answered, to get help, and, if he thinks I'm going to have an attack while I am in the bath, he'll pull the plug out.
>
> (2011, p. 196)

How do salmon 'return home' to their place of birth, or birds know to migrate? Young birds even migrate to areas they have never been before without being led by birds who 'know the way'. Many insects do the same. Jung sees it as follows:

> Viewed from the psychological standpoint, extra-sensory perception appears as a manifestation of the *collective unconscious*. This particular psyche behaves as if it were *one* and not as if it were split up into many individuals. . . . It manifests itself therefore not only in human beings but also at the same time in animals and even in physical circumstances. . . . I call these latter phenomena the synchronicity of archetypal events. For instance, I walk with a woman patient in a wood. She tells me about the first dream in her life that had made an everlasting impression upon her. She had seen a spectral fox coming down the stairs in her parental home. At this moment a real fox comes out of the trees not 40 yards away and walks quietly on the path ahead of us for several minutes. The animal behaves as if it were a partner in the human situation. (One fact is no fact, but when you have seen many, you begin to sit up.)
>
> (Jung, Letters Vol. 1 dated November 1945, p. 395)

Sheldrake has written about this. Picking up on an owner's return home cannot be explained only by the animal's ability to hear or smell more acutely than us. Nor is it connected to regular routine or recognising the owner's car. Sheldrake's book is filled with examples of connections between owner and animal, as well as of the healing potential of living with them. An example:

> Some people who suffer with migraines have dogs that come and lie with them while they are suffering. Frau R. Huber of Horgen, Switzerland, found that her dog Nero also knew on which side of her head she had the migraine. "If it was the right side he excitedly and vigorously licked my right eye and my right forehead with a low whimper. If the pain was on the left he did the same on the other side. It was like a massage".
>
> (Sheldrake, 2011, p. 77)

Sheldrake sees these phenomena in terms of what he calls Morphic Resonance, which he visualises as being like an elastic band connecting creatures and objects (of which more anon).

A different sort of example of animals helping us to heal – or guiding humans – is the case of the Asian tsunami in 2004. Many people noticed that animals were behaving in unusual ways. As well as being able to detect internal dangers, they are able to recognise imminent danger in the outside world. In an unpublished talk (2013), Antonia Boll (a British Jungian Analyst) cites the case of the Asian Tsunami:

> [My colleague] was sailing in the Indian Ocean at the time of the (2004) Box-ing Day Asian Tsunami. She and others in her group awoke to find the sea and sky an unusual colour of purple. Thinking nothing of it they took a small launch and headed for an island that turned out to be 50 nautical miles from the epi-centre of the quake. As they were settling down to explore, they noticed it was impossible to place their mats on the beach. Everywhere they looked, hordes of ants were running in helter-skelter directions. Sensing the ants knew some-thing they did not, [my colleague]'s group read their activity as a sign to move away from the beach as quickly as possible. The only way was uphill. The ants had sensed danger in the earth's tremors and were moving away from the beach. [My colleague] had a lucky escape! At the time of the Tsunami, I also heard about groups of elephants who were seen heading for higher ground. The humans who followed their example and fled inland were saved . . . Anthropol-ogists expected the aboriginal population of the Andaman Islands to be wiped out but they heeded their folklore tradition which speaks of a "huge shaking of ground followed by a high wall of water". The Onge aboriginals all survived.

Shamanic animals

The term 'shaman' originates in Siberia, although there are shamans in most native societies (see Chapter 9 of this volume). Shamans are spirit healers and guide by communicating with the spirit worlds, usually with the assistance of ani-mal helpers. Romanian historian of religion, Mircea Eliade (1907–1986), leading thinker in his field, says:

> the relations between the shaman . . . and animals are spiritual in nature and of a mystical intensity that a modern, desacralized mentality finds it difficult to imagine . . . donning the skin of an animal was becoming that animal, feel-ing himself transformed into an animal.

> (1964, p. 459)

A Lakota medicine woman speaks of her journey:

> When I was having difficulty, I was told Coyote would come and turn me in my path. He is cunning and so I came to a fork in the road – to the left was my old ways, but to the right was the road to health, to spirit, body, nature.

There was sun, dew on the grass, an early morning chill, reflection of sun on dew creating a rainbow effect. . . . You feel [the buffalo's] presence, you smell him, feel his warmth, his coat, feel everything – when he shifts hoofs, you hear that. You experience him huge behind you, all around you. Your own body opens up, your senses respond, it is tactile, smell, you feel it. It is nonverbal. When Elk comes, he has a different smell, his coat is coarse. You feel the warmth through his nostrils. Sometimes he will stroke my hair, sometimes pat my head, relaxing me, meaning that I am too hard on myself. Or he might tap me next to my heart meaning that I have some work to do with my feeling. Or he might give me a handful of herbs which I can smell. Animals teach humour saying "You're worrying too much", implying you're OK as you are. If Eagle comes, it has to do with trust. I have to trust to fly with him.

(quoted in Russack, 2002, p. 47)

For the Oglala Sioux, spiritual questing is sometimes done through dreams to ready the quester for the message. It can also work the other way around: dreams can prepare the seeker to envision. Black Elk, a Lakota holy man, asked writer and scholar Joseph Epes Brown (1920–2000) to create a written account of the Sioux nation's spiritual legacy which was published in 1992 (revised in 1997). Black Elk said:

I was taken away from this world into a vast tipi, which seemed to be as large as the World itself, and painted on the inside were every kind of "four-legged being", "winged-being", and all the "crawling people". The peoples that were here in that [sweat] lodge, they talked to me, just as I am talking to you.

(Epes Brown 1997, p. 50)

Epes Brown tells us there are patterns or combinations of patterns to the dream or vision experiences which arise:

association of the animal or bird spirit-form with the powers of the four directions . . . and the apparition frequently involving four beings, or sets or multiples of four; the appearance of the vision form being heralded by audible messages, often in the form of questions or instructions to the lamenter; and the seeker commonly being transported either to another terrestrial location such as a beaver's lodge, or away from this world.

(ibid., p. 51)

Jung would have delighted at the appearance of quarternities in these images (since he found them to be so significant). They also appear in Oglala Sioux ceremonials:

Sixteen horses were secured. Four of these horses were black, symbolizing the West; four were white, for the North; four were sorrels, for the East; and four were buckskins for the South.

(ibid., p. 58)

A contemporary shaman in the UK gives us an insight into her practice:

> Rather than wearing a specific outfit or using objects imbued with meaning to access animal spirits, my typical approach is to subtly shift my attention to the other realm – to listen with a different ear and look with a different eye. When I ask for help, it may come in animal form directing attention to something important that has been missed. It will occur within seconds.
>
> Gratitude is <u>very</u> important. There is the idea of a respectful trade. I may be asked to realise something in this realm that is important to the spiritual realm, in return for the guidance offered.
>
> (J. Woolliscroft, personal communication April 2018)

The shaman then addresses the environment which needs to be created for such ritual practices:

> Maintaining a pleasant and hygienic space is important for all creatures, and yet I am not one for extensive space clearing as I do not experience animal spirits as having negative energies. In contemporary shamanism a snake is not "evil". It appears to us in order to pass on an important message, intrinsic to its nature. We are invited to pay attention to the ruthlessness of the magpie, the delicate footing of the deer, the ability to strike like the snake. Even parasites, which typically evoke horror, are a natural element that tests the health of a system. A person who is visited by a numinous cockroach may have qualities of resilience to survive events that would destroy others. The closest that contemporary shamanism comes to negative energy in the world of animal spirits is that of blocked natural processes, any setting that blocks the expression of an animal's true nature, so, for example, any creature forced to live in a cage.
>
> (ibid.)

The shaman then addresses the ethical attitude required to undertake such work:

> If there is an ethic conveyed by a shamanic animal presence it is often this. . .
> "Be true to your nature and – if you are struggling – I will lend you my gifts until you find your way".
>
> (ibid.)

In dreams and visions, animals and humans can turn into one another. In myth, they can be both at the same time; think of the centaur. Or they may transmute into plant form to guide the shaman towards a particular medicinal cure using plants or herbs. The Oglala Sioux believed the presence of the spirit is contained in every being – even the tiniest ant – as in the hazardous situation of the Tsunami.

Therapy with animals

The ecopsychology movement is concerned with man's despoliation of the earth and our disconnection from the more-than-human world. Buddhist author and eco activist Joanna Macy is discussed elsewhere in this book (chapter 2). Others have been developing therapeutic approaches to working directly with animals. Therapy with dogs is becoming established. A study analysed therapy with dogs in a university setting:

> sessions had strong immediate benefits, significantly reducing stress and increasing happiness and energy levels. In addition, participants in the experimental group reported a greater improvement in negative affect, perceived social support, and perceived stress.
>
> (Ward-Griffin et al., 2018, pp. 1–6)

Animal assisted psychotherapy (AAT) and equine therapy are both rapidly growing fields where therapists work with the animals in a session. Though I cannot personally vouch for them, there are courses run by the Society for Companion Animal Studies for AAT. Equine therapy can be explored via organisations such as LEAP and Eagala, the leading international non-profit associations for professionals incorporating horses to address mental health and personal development needs (see web links in References).

In equine therapy, the horse is seen as a therapist, or co-therapist. Horses have remarkable intuition and hone in on a site of trauma with precision. They have an uncanny ability to tune in to what is going on with humans and have shown extraordinary powers to heal and help people who turn to them in times of need. In France, horses are being used to advantage with Alzheimer's patients where a scheme has been set up to visit patients at residential homes. In one case:

> [a]n 87-year-old woman who had previously spent her days curled up in a foetal position in bed is now eating meals with other residents and even singing in her native Nicois.
>
> (*The Connexion,* 2018)

My colleague, Jungian analyst Joanne Spilios in New England, describes her therapeutic work together with horses:

> The work involves transference, projection, the collective and personal unconscious, shadow, enlightenment, the Self and the process of individuation. With the horse acting as guide and co-therapist, the unconscious becomes conscious so subtly that even the therapist of any discipline and/ or client/patient opposed to the analytical approach cannot refuse to become enlightened.
>
> (2017, p. 229)

She finds that not only does the horse itself have a curative effect, but seeing them in a herd evokes family dynamics which can be helpful to the therapy. She observes people have to be authentic with horses if they wish to engage or the horse simply will not cooperate (ibid., p. 255). She quotes a participant:

> *Horses are natural Yogis. Because of their energetic state, just by being in their proximity, they entrain our nervous system to balance towards the parasympathetic nervous system, so it lowers our blood pressure, lowers our arousal level. And this happens just by being in their proximity without doing anything.*

(ibid., p. 254; italics in original)

Professor of neuroscience Jaak Panksepp (1943–2017) demonstrated all mammals share 'core emotional processes'. He is known for experiments with tickling rats which showed they laugh! His *Affective Neuroscience: The Foundations of Human and Animal Emotions* (1998) is a classic in the field.

Some therapists have their cats or dogs in the therapy space. The animals know when the patient is in need of comfort and affection and intuitively respond in a way the therapist cannot because of the wisdom of abstaining from too much physical closeness as it could potentially be seductive or misleading. The simple closeness of an animal can be deeply healing. It puts you in touch with your own instinctual/animal nature. An animal showing affection dispels negative feelings about yourself. An anecdotal example is of my favourite 'animal therapist' Alfie (a dog) who would wait at the front door for a particular patient. Even when sessions were rearranged and happened at unfamiliar times, or the patient arrived in a taxi (no recognisable/familiar sound) he would be loyally waiting there for her arrival. Freud allowed his beloved chow, Jo-Fi, to sit with him in sessions. When the patient was calm, the dog would sit near them. When anxious, further away. Jo-Fi even acted as timekeeper and would get up and stretch, heading for the door at the end of the analytic hour (Small, 2007, p. 22).

Although animals and nature have been discussed in a general context, it is useful to think about mechanisms behind some analytic theories. To quote Samuels:

> We don't need to ask why projections travel, because they don't travel – the individuals concerned are already linked.

(1989, p. 169)

This relates to how images intuitively arise in the analyst's mind's eye which often have uncanny acuity in terms of what the patient/analysand is contemplating; and how dreams are shared (Williams, 2019). We are linked.

Merger is usually seen in pejorative terms signifying an inability to separate and become independent. Freud called this an oceanic state. Erich Neumann (1905–1960), a contemporary of Jung, called it Uroboric, a state of primordial chaos (Neumann, 1955, p. 18). Perhaps such connections to earlier developmental states are not pathological. It is the ability to melt those boundaries which enables the shaman to enter a trance state. Naturally, having the ego strength to return to oneself is equally vital.

The theory behind the connections I discuss relate to the idea of a unitary reality where there is no separation between nature and psyche and matter. I will now sketch out various theories which can be used to understand the idea of non-separation. We have probably all experienced the sensation of looking at someone from behind only for them to turn, feeling the gaze on the back of their head. Or we contact someone at the exact moment when they are thinking about us. These ideas are offered as possible avenues for further exploration, and can only be surveyed in brief here.

Systems conceptualising energy fields

Psychoid Unconscious

Jung first suggested there was a 'psychoid' unconscious completely inaccessible to consciousness in 1946. Jung linked this Psychoid Unconscious to the *unus mundus*, or 'one world', where everything is invisibly connected on a subtle level. *Unus mundus* was a term he adopted from 16th-century alchemist Gerhard Dorn, a student of German Renaissance astrologer Paracelsus (1493–1541). The *unus mundus* relates to the unitary nature of reality beyond the Cartesian split between mind and body. Jung calls the *unus mundus* a 'metaphysical speculation' (CW 11, para. 660), a theory which help us think about something intangible:

> The psychoid archetype / unconscious has a tendency to behave as though it were not located in one person but were active in the whole environment. The fact or situation is transmitted in most cases through a subliminal perception of the affect it produces. . . . As soon as the dialogue between two people touches on something fundamental, essential, and numinous, and a certain rapport is felt, it gives rise to a phenomenon which Lévy-Bruhl fittingly called *participation mystique*. It is an unconscious identity in which two individual psychic spheres interpenetrate to such a degree that it is impossible to say what belongs to whom.
>
> (CW 10, para. 851–2)

Jung located the psychoid archetype:

> beyond the psychic sphere, analogous to the position of the physiological instinct, which is immediately rooted in the stuff of the organism and . . . forms the bridge to matter in general.
>
> (CW 8, para. 420)

It is where the link between psyche and matter is revealed not only as two sides of a coin (a revolutionary enough idea), but as the same thing. Main encapsulates the complexity when he says the important thing is the psychoid is simply not physical plus psychic because the categories become meaningless. He suggests the concepts:

> of psyche and matter and space and time merge into a psychophysical space-time continuum. . . . To express this ambivalent nature – at once psychic and physical yet neither because beyond both – [Jung] was led to coin the term "psychoid".
>
> (1997, p. 36)

In 'Psychological Commentary on the Tibetan Book of the Great Liberation', Jung contrasts Eastern and Western modes of thinking, and elaborates (before formally theorising the Psychoid Unconscious) the linked nature of 'one mind':

> The statement "Nor is one's own mind separable from other minds" is another way of expressing the fact of "all contamination". Since all distinctions vanish in the unconscious condition, it is only logical that the distinction between separate minds should disappear too.
>
> (CW 11: par. 817)

Implicate Order

Theoretical physicist David Bohm (1917–1992), one of the most innovative theoretical physicists and a protegé of Albert Einstein, postulated an *Implicate Order* whereby 'it is possible to comprehend both cosmos and consciousness as a single unbroken totality' (1980/2002, p. 219). He refers to the Ancient Greeks – the school of Parmenides and Zeno, who suggested space is a plenum; it is 'matter' rather than empty space. Bohm writes:

> [w]hat we perceive through the senses as empty space is actually the plenum, which is the ground for the existence of everything, including ourselves. The things that appear to our senses are derivative forms and their true meaning can be seen only when we consider the plenum, in which they are generated and sustained, and into which they must ultimately vanish.
>
> (ibid., p. 243)

The sea of energy belonging in the realm of the plenum – which constitutes the plenum – is a holographic, multidimensional Implicate Order. When you think about space and air and the whole cosmos as being made up of energy in this way, it becomes possible to see how communications may be transmitted as vibrations between us; between us and animals; and between us, the earth and all of nature. Consciousness and materiality are connected. Bohm puts it more strongly:

> [i]t will be ultimately misleading and indeed wrong to suppose . . . each human being is an independent actuality who interacts with other human beings and with nature. Rather, all these are projections of a single totality.
>
> (ibid., p. 266)

Bohm shows us the undivided wholeness of our natural state: the earth, the beings, the energy and everything which makes up the universe as one. This explains the butterfly effect from chaos theory whereby a tiny movement sends ripples out through the entire cosmos, and it explains how we can unconsciously communicate between ourselves (see Jung's Gate diagram, CW 16, para. 422), and with animals and our [unconscious] impact on the earth.

Bohm refers to his central thesis as an 'unbroken wholeness of the totality of existence as an undivided flowing movement without borders' (ibid., p. 218). This has resonance as a political idea in thinking about immigration policies. What would life be like if we took Bohm's findings seriously as a collective? What would this mean in terms of the 'other'? How can there be an 'other' when we are all connected?

Brian Josephson, Nobel prize-winning physicist from Cambridge University, believes that Bohm's Implicate Order 'may someday even lead to the inclusion of God or Mind within the framework of science' (Talbot, 1991, p. 54).

Akashic Field

The Akashic Field was conceptualised by Hungarian former professor of philosophy Ervin Laszlo. Akasha is the Sanskrit word for ether, the cardinal element alongside the other four: earth, fire, water, air. Laszlo describes the universe using similar terms to Bohm:

> *The most fundamental element of reality is the quantum vacuum, the energy- and in-formation-filled plenum that underlies, generates, and interacts with our universe.*
>
> (2004, p. 103; italics in original)

He turns to Hindu and Chinese cosmologies which:

> have always maintained that the things and beings that exist in the world are a concretization or distillation of the basic energy of the cosmos. . . . The physical world is a reflection of energy vibrations from more subtle worlds that, in turn, are reflections of still more subtle energy fields.
>
> (ibid.)

He refines his argument:

> In Indian philosophy the ultimate end of the physical world is a return to Aka-
> sha, its original subtle-energy womb. At the end of time as we know it, the
> almost infinitely varied things and forms of the manifest world dissolve into
> formlessness, living beings exist in a state of pure potentiality, and dynamic
> functions condense into static stillness. In Akasha, all attributes of the mani-
> fest world merge into a state that is beyond attributes: the state of *Brahman*.
> (ibid., pp. 103–4; italics in original)

Brahman is a Sanskrit word for God, or ultimate reality beyond splits and duality.
It is these subtle energies which give us access to the connections with animals
and the other-than-human world and which challenges Western rationalism and
scientific materialism.

Morphic Resonance

Similar notions of interconnectedness and wholeness are found in the ideas of
Rupert Sheldrake writing from the biologist's perspective about 'Morphic Resonance' (2009). He points out that modern science recognises interconnection in
the effects of gravity and magnetic fields on living things. Tides are influenced
by the moon. We are accustomed to using microwaves, lasers, radiation and wifi.
Sheldrake tells us physicists speculate everything in the universe is interconnected
through quantum non-locality:

> Once two particles have interacted with one another they remain linked in
> some way, effectively parts of the same indivisible system. This property of
> 'non-locality' has sweeping implications. We can think of the Universe as a
> vast network of interacting particles, and each linkage binds the participating
> particles into a single quantum system.
> (Paul Davies and John Gribbin, quoted in Sheldrake, 2011, p. 236)

This helps us recognise our interdependency as a human species and encourages
us to respect our links with animals and nature. It also makes us think about the
struggles people experience in 'letting go' after relationship break ups or death.
Does the link ever come to an end? The grieving process entails cutting the binds
metaphorically, but perhaps the link is deeper than we know. When souls connect,
there is a feeling they remain connected throughout time.

The term Morphic Resonance derives from what biologists call morphogenetic
fields which shape developing organisms. They are blueprints underlying the
form of the growing organism (Sheldrake, 2011, p. 258). Morphic fields are self-
organising regions of influence in space and time which guide the system towards
a goal. They contain a memory which acts as a resonance, or Morphic Resonance

(ibid., pp. 260–1). Sheldrake suggests this system implies a non-locality not currently recognised in mainstream science:

> [i]t may turn out to be related to the non-locality or non-separability that is an integral part of quantum theory, implying connections or correlations at a distance undreamt of by classical physics. Albert Einstein found the idea of "spooky action at a distance" implied by quantum theory deeply distasteful; but his worst fears have yet to come true.
>
> (ibid., p. 262)

Holographic Universe

The originators of the idea of the Holographic Universe are David Bohm (see earlier in this chapter) and Karl Pribram, neurophysiologist from Stanford University and author of *Languages of the Brain* (1971) who worked separately but came to similar views. American writer Michael Talbot (1953–1992), in his project to incorporate spirituality and science, surveys these pioneering thinkers with the benefit of modern research. Talbot goes back to the originator of quantum physics, Nobel prize-winning Danish physicist Niels Bohr (1885–1962), who originated the idea the observer influences research. Bohr went so far as to say subatomic particles only come into existence with the arrival of the observer. An image which might help understand what he means is this: imagine if every time we put our foot down to walk, the ground beneath us manifests to support us.

Some dismiss Jung's notion that stones have soul as 'hippy dippy'. If that is true, then Bohm must be, too. In his discussion of Bohm, Talbot tells us Bohm believes:

> dividing the universe up into living and nonliving things also has no meaning. Animate and inanimate matter are inseparably interwoven. . . . Even a rock is in some ways alive, . . . for life and intelligence are present not only in all of matter, but in "energy", "space", "time", "the fabric of the entire universe", and everything else we abstract out of the holomovement and mistakenly view as separate things.
>
> (1991, p. 50)

A hologram is a recording of a light image which can then be replayed, or reproduced, in 3D. Each element contains the seed of the whole so that: 'the form and structure of the entire object may be said to be *enfolded* within each region of the photographic record' (Bohm, 1980/2002, p. 225; italics in original). Implicate means to enfold, so Bohm's premise is that everything is enfolded into everything else.

An image which might help to visualise the idea of the universe as hologram is that of Indra's net from Indian and Chinese Buddhist philosophy, quoted by American Jungian analyst Joseph Cambray:

> In the heaven of the great god Indra is said to be a vast and shimmering net, finer than a spider's web, stretching to the outermost reaches of space. Strung at each intersection of its diaphanous threads is a reflecting jewel. Since the net is infinite in extent, the jewels are infinite in number. In the glistening surface of each jewel is reflected all the other jewels, even those in the furthest corner of the heavens. In each reflection, again are reflected all the infinitely many other jewels, so that by this process, reflections of reflections continue without end.
>
> (2009, p. 44)

Borderland consciousness

Contemporary Jungian analyst Jerome Bernstein from Sante Fe, New Mexico, has conceptualised what he calls 'Borderland' consciousness. Having originally worked for the federal government in the United States, Bernstein was invited by the Navajo nation to consult on creating systems to restore the tribe's culture, language and dignity. He realised that his heart belonged with this community and gradually moved there, becoming adept in their languages and healing traditions. This opened him to the undivided wholeness being discussed. Bernstein sees borderland (not border*line*) personalities as being gifted to experience the split off aspects of life:

> Borderland people *personally* experience, and must live out, the split from nature on which the western ego . . . has been built. They feel (not feel *about*) the extinction of species; they feel (not feel about) the plight of animals that are no longer permitted to live by their own instincts. . . . Such people are highly intuitive. Many . . . are psychic. . . . They are deeply feeling, sometimes to such a degree . . . that seem irrational to them. Virtually all of them are highly sensitive on a bodily level. They experience the rape of the land in their bodies.
>
> (2005, p. 8; italics in original)

Bernstein quotes Al Gore, whose journey through the humiliation of losing in the US presidential election of 2000 is a beacon to us all. He shows how losing is often winning:

> We have assumed that our lives need no real connection to the natural world, that our minds are separate from our bodies, and that as dismembered intellects we can manipulate the world in any way we choose. Precisely because

we feel no connection to the physical world, we trivialize the consequences of our actions.

(Gore, 2000, quoted in Bernstein, 2005, p. 33)

This is important not only in terms of our connection to animals, but to the earth and nature. They are an invitation to us to open to the energies embodied by our fellow creatures on this earth to understand our place and how we can be of service as well as how we can damage and despoil.

The work of Bohm, Sheldrake, Talbot and Laszlo, as well as others mentioned in this chapter, provide a scientific foundation to Jung's postulation of a Psychoid Unconscious where psyche and matter meet. Science has caught up with Jung's intuition.

Two other writers should be mentioned by way of signposts for further exploration. Fritjof Capra's famous *The Tao of Physics: An Exploration of the Parallels between Modern Physics and Eastern Mysticism* (1975/1991) was followed by his 2014 volume *The Systems View of Life: A Unifying Vision*, co-authored with Pier Luigi Luisi (Capra and Luisi, 2014), and Lawrence Le Shan's *The Medium, the Mystic, and the Physicist: Toward a General Theory of the Paranormal* (2003). Reading the physicists mentioned here can be mind-blowing in its scope. I find that it can be difficult to understand, but it touches on the deepest levels of consciousness and existence.

The connections made in this chapter between man, nature, animals and the universe have significance for the healing of ourselves and the *Anima Mundi*, the soul of the world. To rediscover the natural world as beings – not objects – is truly vital. Writing this chapter has been an exercise in holding the sacred and the grounded in balance to ensure the claims I have made are on solid ground. This to me is a metaphor for life. Don't we have to manage these aspects of living each day? I hope that I have shown how fundamental animal energies are and ways they can be honoured and contacted in daily life, helping us give due credence – and love – to the more-than-human world which surrounds us and of which we are a mere part.

Wishing you the wisdom of the owl, the perspicacity of the cat, the instincts of wild beings, the cunning of the fox, the patience of the snail, and the intelligence of dolphins as you make your way through this book.

Acknowledgements

Some of the material in this chapter was originally published in *C.G. Jung: The Basics* (Routledge, 2019) by Ruth Williams and appears here with kind permission of Routledge. I am grateful to Moira Duckworth for drawing my attention to the article from *The Connexion* quoted in this chapter. My thanks to Antonia Boll for permission to quote from her unpublished paper. And I am grateful, as ever, to Professor Andrew Samuels for his support and encouragement in writing this chapter.

References

Bernstein, J.S. (2005). *Living in the borderland: The evolution of consciousness and the challenge of healing trauma.* London and New York: Routledge.

Bohm, D. (1980/2002). *Wholeness and the implicate order.* London and New York: Routledge Classics.

Boll, A. (2013). *Listening to the earth's tremors: The helpful animal as catalyst in the journey towards wholeness.* Unpublished talk.

Brown, J.E. (1997). *Animals of the soul: Sacred animals of the Oglala Sioux.* Rockport, TX: Element Books.

Cambray, J. (2009). *Synchronicity: Nature & psyche in an interconnected universe.* College Station, TX: Texas A & M Press.

Capra, F. (1975/1991). *The Tao of physics: An exploration of the parallels between modern physics and eastern mysticism.* London: Flamingo.

Capra, F. and Luisi, P.L. (2014). *The systems view of life: A unifying vision.* Cambridge and New York: Cambridge University Press.

The Connexion. (2018). *Patients' eyes light up as horse pays a visit.* News in Brief, Apr. 11.

Eliade, M. (1964). *Shamanism*, Bollingen Series LXXVI. Princeton, NJ: Princeton University Press.

Hillman, J. (1997). *Dream animals*, Writings by J. Hillman. Paintings by M. McLean. San Francisco, CA: Chronicle Books.

———. (2008). *Animal presences. Uniform edition of the writings of James Hillman,* Vol. 9. Putnam, CT: Spring Publications.

Jung, C.G. (1936/1989). *Nietzsche's Zarathustra: Notes on the seminar given in 1934–1939 by C.G. Jung.* ed. by J.L. Jarrett. Vol. 2. London: Routledge.

———. (1945). Answers to Rhine's questions, dated November 1945. In: *C.G. Letters Vol.1 1906–1950*, selected and ed. by G. Adler in collaboration with A. Jaffé, trans. R.F.C. Hull, in two volumes. London: Routledge and Kegan Paul.

———. (1953–77). *Except where indicated, references are by volume and paragraph number to the collected works of C. G. Jung.* 20 vol., ed. by H. Read, M. Fordham, and G. Adler, trans. by R.F.C. Hull. London: Routledge and Princeton, NJ: Princeton University Press.

Laszlo, E. (2004). *Science and the Akashic field: An integral theory of everything.* Rochester, VT: Inner Traditions.

Le Shan, L. (2003). *The medium, the mystic, and the physicist: Toward a general theory of the paranormal.* New York: Allworth Press.

Main, R. (1997). *Jung on synchronicity and the paranormal.* Princeton, NJ: Princeton University Press.

Neumann, E. (1955). *The great mother.* London: Routledge and Kegan Paul.

Panksepp, J. (1998). *Affective neuroscience: The foundations of human and animal emotions.* Oxford and New York: Oxford University Press.

Pribram, K.H. (1971). *Languages of the brain: Experimental paradoxes and principles of neuropsychology Prentice-Hall series in experimental psychology.* Upper Saddle River, NJ: Prentice Hall.

Pullman, P. (1995). *Northern lights.* London: Scholastic Children's Books.

———. (1997). *The subtle knife.* London: Scholastic Children's Books.

———. (2000). *The amber spyglass.* London: Scholastic Children's Books.

Russack, N. (2002). *Animal guides: In life, myth and dreams.* Toronto: Inner City Books.

Samuels, A. (1989). *The plural psyche: Personality, morality & the father*. London and New York: Routledge.

———. (1993). *The political psyche*. London and New York: Routledge.

Sheldrake, R. (2009). *Morphic resonance: The nature of formative causation*. Rochester, VT: Park Street Press.

———. (2011). *Dogs that know when their owners are coming home: The unexplained power of animals*. London: Arrow Books.

Small, S. (2007). *100 dogs who saved civilization*. Philadelphia, PA: Quirk Books.

Spilios, J. (2017). The curative factors of equine-assisted psychotherapy. In: *The association of jungian analysts 40th anniversary Festschrift*. London: AJA.

Talbot, M. (1991). *The holographic universe*. London: Harper Collins.

Ward-Griffin, E., Klaiber, P., Collins, H.K., Owens, R.L., Coren, S. and Chen, F.S. (2018). Petting away pre exam stress: The effect of therapy dog sessions on student well-being. *Stress and Health*. https://doi.org/10.1002/smi.2804.

Williams, R. (2017). The function and meaning of the image of "the witch" in the individuation process of women. In: *The association of Jungian analysts 40th anniversary festschrift*. London: AJA.

———. (2019). *C.G. Jung: The basics*. London and New York: Routledge.

Websites

Equine Assisted Growth & Learning Association (Eagala): www.eagala.org.
Society for Companion Animal Studies: www.scas.org.uk/.
UK LEAP: www.leapequine.com/.

Albatross

No shadow falls
On ocean skin
Over lost bones
Lost love lost dreams
On simple wings
He rides the winds
Rough smooth dark light
The turning waters
Frame his flight
On my heart
No shadow falls

Grant Clifford

Part IV

Myths and futures

Chapter 11

Time, intuition and imagination

Dale Mathers

> The idea of the future, pregnant with an infinity of possibilities, is thus more
> fruitful than the future itself, and this is why we find more charm in hope than
> possession, in dreams rather than reality.
>
> (Bergson, 1912/2018, p. 10)

What is change?

There is so much 'gloom and doom' around climate change, I felt in need of a
fresh approach. So, before writing this, I asked for a dream. . .

> In an old library, an older friend, the late George Blaker, sits in a leather chair,
> with an open book on his desk. He smiles and says, "Re-read 'the Imprisoned
> Splendour,' and you will find the philosophy you need".

In 1975, whilst I was a young medical student, George lent me this book.

Its author, Raynor Johnson, was Professor of Physics at the University of Mel-
bourne. The 'Imprisoned Splendour' is the name he gave to the spirit trapped
in the material world (Johnson, 1953). Here, 'spirit' means 'the Eternal soul' as
well as 'spirit' – as in spiritualism. Johnson gives descriptive accounts of many
paranormal experiences: clairvoyance, mediumship, ghosts, precognition, teleki-
nesis. Jung researched parapsychology for his doctorate, writing about his young
cousin, Helene Preiswerk, a self-proclaimed trance medium (CW 1, para. 1–165).
His convincement of the reality of the paranormal contributed to his split from
Freud (Main, 1997, pp 2–5).

Jung became fascinated by alchemy, a 'great grandparent' to natural philosophy
and science. Alchemists supposed 'the Divine spirit' was imprisoned in matter.
So, alchemy as a metaphor is an insight from this dream. Metaphor is part of the
insight, also – metaphors are symbols for transformation. For me, a library symbolises transformation: books are distillations of thoughts and feelings, where the
'gold' of new thinking may be hidden.

In this chapter, I look at how understanding time, using intuition and imagination, can help us face into climate change and how a perennial philosophical

problem – the distinction between spiritual and material explanations about 'being in the world' – entangles how we navigate time. Time navigation is a skill. Its use depends on our occupation and our culture. Musicians are highly tuned to time. Analysts are good at knowing how long fifty minutes is. I'll look at the 'behind, below and above' of time, as its perception (or lack of it) is a contributing cause of climate change.

Naturally, we simply cannot experience time beyond 'our lifetime' – about seventy-five years. We may know a great grandparent; we may meet a great grandchild. Three generations behind and before gives a transgenerational span of about a hundred and fifty years. For a beech tree, a transgenerational span is about 1,200 years; for a bacteria, maybe twelve hours. So, actions with consequences in a future we can never know obviously can't be felt-and-lived experiences; they can only be perceived by intuition or imagination.

Our planet and its many ecosystems survived several Ice Ages. Now it's over-heating. Climate always changes, and change is usually natural. Whether for a galaxy or a planet, change is a result of existing in time. We don't usually think about time in the abstract; we live in it as a fish lives in water, swim in its flow from past to present to future, and know that, for each of us, 'our time' will end. Mostly, we live in denial of our own death, to make our personal world feel safe. But denial of time's flow can never work. Time catches up with us.

Time also 'permits' free will. Without it, actions cannot have consequences. Temporal flow creates causality. In the West, this is a basic metaphysical concept: 'meta' is Greek for 'behind, below or above'; metaphysics is the 'behind, below and above' of physics (formerly called 'natural philosophy'). In the East, the idea that actions have consequences isn't a metaphysical argument; it is an everyday fact called karma. As a Buddhist, I'd say all our experiences arise from karma. Climate change is our karma. This hardly seems worth saying, except when denial becomes a political change-management technique. Denial, like splitting and projective identification (analysts call these 'the primitive defences of the Self'), works against reality testing. When you remove reality, you remove freedom of choice.

We don't have to 'save the planet'. Our planet, as a planet, is perfectly safe – until the sun expands in about three billion years. The ecosystem is in danger, and this sense of danger invades our inner worlds, our minds, creating shame and guilt, evoking 'primitive defences'. Our minds are likewise complex ecosystems – neural forests, each with over ten billion interconnected trees. Through an unconscious sharing percepts and concepts, we connect to others by the collective unconscious. This natural network uses intuition and imagination to find new choices, to adapt to new realities. We have no choice about the reality of climate change; we have many choices about *how* we face it. We could become angry, grief-stricken or both. An early description of eco-grief is given by American hospital chaplain and ecologist Phyllis Windle in 'Ecopsychology' (Windle, 1995, pp. 136–46): it feels the same as any other bereavement.

Floating on a raft of denial, rage, guilt and shame, eco-grief arises from our inability to experience time beyond 'our lifetime'. Loss can be painful beyond imagining. Yet, till recently, we simply couldn't imagine where our old trainers or plastic bags went. Now we can, and do, and despair. So, to start finding hope, I'll tell you about the Scientific and Medical Network.

The Scientific and Medical Network

Amplifying the dream further, in the early 1970s, George Blaker, a senior British civil servant, co-founded the Scientific and Medical Network with Dr. Patrick Shackleton, then Dean of Southampton Medical School. Once, Patrick gave me a lift, and our conversation was life-changing. We spoke about the 'Imprisoned Splendour', sharing concerns about the exclusion of the spiritual, in any form, from our discipline and its replacement by scientific (and medical) materialism – fundamentalism in a white coat. As a young member of the Student Network, I watched John Davy (science correspondent of the *Observer* newspaper) hold up a micrometer and ask, 'Can you measure happiness by using this to measure a smile?' He introduced the distinction between the quantifiable – scientific – and the unquantifiable – inner – experience.

This distinction was made by the French philosopher Henri Bergson. Distinguishing between quantifiable time (how long it will take for the ice caps to vanish) and qualitative time (how deeply will we grieve after the ice caps vanish) helps us be aware of both time-bound and time-free experience. This discrimination helps in adapting to change. Perhaps, as other writers in this book suggest, denying 'the spiritual' fosters abusive and neglectful attitudes to each other and all living beings. But materialism isn't a synonym for greed. In philosophy, materialism means imagining one 'substance' makes everything: consciousness, thought and feeling arise *only* from material interactions. As a Buddhist, I might say, 'arise from material interactions'; the word '*only*' is the problem – as in 'there is *only* the material world'.

The idea 'everything is quantifiable' (measurable) and *only* the quantifiable exists belongs to scientific monism. Its opposite, spiritual monism, isn't synonymous with 'good'. There are many 'self-styled' gurus and more-than-doubtful religious practices, at least as problematic as scientific materialism. They may create 'transcendent defences', like this: 'Climate change? It's God's holy plan! We live in the Tribulation before Judgement Day, when the righteous . . .' and so on – a popular, populist discourse.

Seeing the universe as both spiritual and material is called dualism. It has several forms; some suggest certain things are beyond human understanding yet are knowable to 'a creator God'. The philosophical problem – monism versus dualism – matters for eco-crisis because how we see 'spirit' and 'matter' shapes how we navigate time and whether we retain free will. As 'actions have consequences' and climate change will affect future people more than three generations ahead, we are free to choose not to endanger them. English philosopher Charlotte Wright explains (2018, p. 9) that current people have a moral obligation to future people. She argues there are two ways to approach this:

> The Rights Account – future people have rights which we ought not to violate. These rights mean that we ought to benefit future people in certain ways, or else we are being immoral.
>   And
> The Value Account – we should do that which makes the world go best, and whenever that entails protecting or benefiting future people, we ought to protect or benefit them.

Like Wright, I favour the value account. Measuring 'rights' leads to endless legal arguments. Perhaps 'values' are an 'Imprisoned Splendour', the notion that we share a common humanity? Let's look at a different sort of law: natural philosophy, the project of including the spiritual with the material. Let's step back a hundred years, to Henri Bergson.

Henri Bergson

Henri Bergson was born in Paris in 1859 to a Polish father and an English mother. After a traditional Jewish religious education, he lost faith during adolescence, finding too many contradictions between dogma and experience. Mathematically gifted, at 18 he won a prestigious national prize. His tutor, learning Henri wished to study philosophy reputedly said, 'what a waste!' (Lawlor and Leonard, 2016). But maths includes learning to write equations – propositions describing reality. This Bergson did, tackling the matter/spirit problem, challenging the view of the eighteenth-century Prussian metaphysician Immanuel Kant.

Bergson taught philosophy in Clermont-Ferrand, France, where, in 1889, he completed *Time and Free Will*, winning his doctorate. In this work, he questions Kant's idea that causality 'belongs to God,' who lives outside of time. Bergson suggests our immediate, inner experience is of duration (*la durée*), the flow of felt time, in which there is no sequencing of events. Felt experiences of time are not measurable (see Bergson, 1912/2018, pp 56–77; Guerlac, 2006, pp. 68–70). In 1907, he wrote 'Creative Evolution', which in 1927 won him the Nobel prize. The English philosopher Bertrand Russell (1912; italics in original) said, 'he writes *literature*'; and regarded Bergson as a distinguished *fiction* writer (though he accepted the same prize in 1950). In 1913, Bergson toured America, lecturing on freedom and spirituality to large audiences. The same year he Chaired the Society for Psychical Research (SPR) in London; then, boundaries between psychology and parapsychology were undefined. The 'Imprisoned Splendour' was open to scientific investigation.

Paranormal (spiritual and spiritualist) phenomena may be understood, in part, as temporal disturbances, when 'inner time' experiences are mistaken for 'outer'.

The late Prof. Arthur Ellison (a Network member, and a chair of the SPR) suggests ghosts are felt memories persisting after the person from whom they came has died (1988, pp. 15–25) Ghosts (apparitions) are, if you like, a legacy for future people. They're a link across time: much analysis involves letting ghosts become ancestors.

In 1922, Bergson 'lost' a debate with Albert Einstein on the nature of time: perhaps he didn't grasp the finer points of quantum mechanics; perhaps Einstein didn't grasp the finer points of philosophy. However, Bergson's distinction between 'material time' (science's time 't') and 'spiritual time', says, like quantum mechanics, what you see depends on where you observe – and what you observe *for*. Like Jung, he suggests we can observe the spiritual, the material or both. His argument is against 'quantifying' and for 'qualifying'. We can't measure inner, felt experience: perhaps that's why we can't 'measure' the paranormal? Then, we can't measure 'guilt' either.

A musical analogy may help. Though we stick to 'the time signature', it is the *feeling* brought to the notes – and the space between them – which gives music its soul. Suppose that spiritual experiences, in inner time, are about feeling. In the dream, my late friend George smiled. The feeling was lightness and relief: 'there is an answer!' Knowing we see time in two ways, inner and outer, helped me gain perspective about climate change: 'change has always happened. We survived the Ice Ages . . .'. The dream led me to Bergson, whose influence on Jung is deep; yet, curiously, hardly acknowledged. His name doesn't appear in American philosopher Marilyn Nagy's book on Jung's philosophical influences (1991). Maybe because of his spirituality and spiritualism, Bergson's reputation lay hidden, until rediscovered by, amongst others, French philosopher Gilles Deleuze (1925–1995). He said:

> If philosophy has a positive and direct relation to things, it is only insofar as philosophy claims to grasp the thing itself, according to what it is, in its difference from everything it is not, in other words, in its internal difference.
>
> (Deleuze, 2004, pp. 32–51)

The 'thing in itself' is 'the thing seen intuitively'. Intuition requires a sense of ourselves as beings flowing within time. So, what is time? Is it material, spiritual or both? We face climate change: we know what climate is, we need to understand what *change* is. It is made of time.

Time

Scientific time (the 't' in equations) is measured in seconds, defined in SI units derived from the natural vibrational frequency of caesium-133. In the early twentieth century, the emergent discipline of psychology obsessed about measurement – using micrometers to measure smiles. Bergson saw that this approach ignored felt and lived experience. In *Time and Free Will*, he suggests time is mobile and incomplete. For an individual, inner time can speed up, slow down, stop or go backwards (as in a déjà vu experience).

Outer time is consistent. Inner time wobbles, remember waiting to see the dentist or to take an exam which we've not prepared for. Analysts notice this: a 'fifty-minute hour' may feel like it passes in seconds or drags on for years. Time bends and twists in our dreams: my friend George died many years ago but is alive in my inner world. Dream time is not clock time. Once, we lived almost time-free lives. Hopefully, you remember how time felt when you were a child. First, there is 'now', 'this present moment'. Gradually, we discover telling the time; yet, at 9 years old, the distance between breakfast and bedtime is vast. As we age, time contracts. Grandchildren turn from babies to teenagers in 'next to no time'.

Time perception depends on our perceptual apparatus. In humans, this is complex, involving several overlapping neural networks. Some determine wake/sleep cycles, involving the pineal gland and melatonin. Others, far larger, determine hormone releases and give the felt-sense of time. Yet others give 'time as a memory': 'I read Jackson's book when I was a student . . .'. We can't do 'fine discrimination' of time. We can't see flickering light if the frequency is greater than twenty-five times a second: so, cinema works by flicker fusion; we see separate images as continuous (a solid line) when they are discrete (a broken line, a packet – a quantum). Bergson emphasised the difference between continuous and discrete experience: between the flow of time and a quantum of time.

So, what is inner time? Clearly, not 'clock time' – the 't' in equations. Scientific time is 'discrete'. This is the 'quantum' in quantum physics – broken lines. The unit of quantum time is 'Planck time' (named after German physicist Max Planck), it is 5.4×10 *to the minus 44* of a second. At this scale, time does not flow. We can only imagine this, as does the Italian physicist Carlo Rovelli (2018, pp. 72–85), who describes, at the quantum level, space-time is granular, like rice. We may feel this in the bubbles of collective despair arising from natural disasters, like the Australian bush fires. Suffering has a granularity.

To explore 'felt' time, we need to be in touch with our inner worlds. Duration (*la durée*) is glimpsed in symbols and images, grasped by intuition and imagination. Bergson's friend, American psychology pioneer William James, said Bergson helped him give up logic, 'squarely and irrevocably' as 'reality, life, experience, concreteness, immediacy, use what word you will, exceeds our logic, overflows, and surrounds it' (James, 1909/2015, pp. 111–36). This helps us now because rational, logical 'scientific' arguments fail to convince many of the reality of climate change. Fine. So, we need illogic, unreason . . . and imagination, one of the gifts of childhood's timeless experience. We need dreams, which are time-free.

Ancient Greek representations of time help make this clearer. Chronos (father of the Olympians) is Old Father Time, Saturn, the grim reaper with the scythe and hourglass who measures 'Life-Time'. Kairos, the young god, is 'the opportune moment' – god of chance and change. London Jungian Yvette Weiner (1996) reminds us when we imagine gods, we're looking at metaphors for psychic processes. So, if time is *both* an inner *and* outer experience, *both* Chronos *and* Kairos, how do we navigate? Do we decide inner time is imaginary and scientific time is real? This is not theoretical: the existence of future people depends on deciding whether climate change is real or imaginary. Time perception is crucial to reality testing.

Reality testing and alethiometry

Here is a fictional 'clinical example', taken from Philip Pullman's fantasy trilogy *His Dark Materials*. In *Northern Lights* (1995, p. 79), the young heroine Lyra is given an alethiometer. It looks like an old-style pocket watch, with three large hands and a small one. It measures truth (in Greek αλήθεια, *Alethia*, means truth). You choose three symbols for your question from the thirty-six around the rim, then push the button. The thin hand, driven by 'dust' (elementary particles linked to consciousness), spins round and settles on a symbol suggesting an answer.

Let's ask about climate change. We choose a skull (death), a globe (the planet), and an hourglass (time), and our answer is a tree. What does a tree mean? Trees link above to below – earthy roots to airy leaves – water is moved up them by the fire of the sun – this is an alchemical metaphor – so, plant more trees. Fourteen billion trees (twenty per person) fixes enough carbon to make a difference. Trees symbolise life, a forest symbolises an ecosystem . . . any symbol is an ecosystem of ideas with multiple layers of meaning, as described in Jungian analyst Leopold Stein's classic paper 'What Is a Symbol Supposed to Be?' (1957). Greek messenger boys were given half a broken disc bearing the sender's seal. The recipient proved their identity by joining their half to the messengers'. In Greek, '*Sym*' means 'together' and '*Bolos*' means 'to throw' – symbols 'throw ideas' together to make new ones.

Symbolic truths are open and flexible, with layers and levels of meaning (Mathers, 2001, pp. 191–6). Analysts play with symbols when we work with dreams, helped by our intuition: a sense of what *may* happen. My first analyst supposed 'paranormal events' were attuned intuition, adding, 'the trouble with intuition is we assume it's true; and it is, half the time. The trouble is, we don't know which half'. Intuition relies on trusting the unconscious *may* pull out from sense data the likeliest of a set of future probabilities. To intuit is to be with duration (*la durée*) where there is no causality. Things *just are*.

As mentioned earlier, speculations about space, time and causality are basic to metaphysics, the 'behind, below and above' of natural philosophy. Imagine an oil painting: if metaphysics is the frame and canvas, then philosophy is the picture. Pictures evoke feeling. Science evokes feelings too, sometimes outrage when it contradicts entrenched political positions. We need to look 'behind, below and above' political discourses, and, as the young climate activist Greta Thunberg said at the United Nations, 'look at the science'. But many people don't want to look at the science. Political discourses have a political purpose – to create an illusion of control to benefit of those in power. Challenging their discourse upsets fundamentalists (and fundamentalist parts of our psyche) which deny change: lying is necessary to maintain closed system thinking. What might open system reality testing be like? This is what an analyst encourages. Could it help us test the truth or falsity of a discourse? How would it do this? To move forward, let's step back – this time, to the eighteenth century.

Percepts without concepts are blind

This famous saying is by Kant, the metaphysician: who is sometimes paraphrased as saying 'we perceive reality not as *it is*, but as *we are*'. This is at the heart of our collective problem with 'being in time' and being with climate change. As this is scientifically, materially, real, how can anyone imagine otherwise? Because they don't have, or choose not to have, the appropriate concepts; or, they lie. Science is based on accurate and repeated perceptions, it's a search for truth. Metaphysics questions perceptions: asking 'what's real, and how do we know it?' Useful, to help recognise lies.

If you had an alethiometer, it would be useless without concepts to fit the symbols. What symbols would you choose if you asked about climate change? Your choices depend on your concepts. If you conceive of spiritual reality as an 'Imprisoned Splendor', this would shape your choice. In the Middle Ages, alchemists (early scientists), didn't fuss about the material/spiritual duality. For them, *obviously*, spirit was trapped in matter (a gnostic idea) and it was their noble task to free it! Base metals (like lead) *obviously* contained the spiritual seeds of higher metals, like gold. All you had to do was encourage the metal to grow (see 'Splendor Solis', an illuminated manuscript on the alchemical process; Skinner et al., 2019, pp. 14–6, 146). The 'obvious' is 'the world-as-it appears-to-us'.

Kant argued cause and effect are in the 'world-as-it-appears-to-us', and so are in space and time. They are concepts, not 'things-in-themselves;' and they are 'concepts' we impose on sense data, rather than features of the real world. He thought there is a world outside our minds which is the way it is regardless of our concepts, yet we have no access to what it is like. He suggested causation occurs *only* in the 'world-as-it-appears-to-us', not in an underlying mind-independent reality. It's not clear Kant consistently thought causation is always *only* part of 'world-as-it-appears-to-us' and not a 'thing-in-itself', because he thought 'things-in-themselves' are the source of our sensory experiences. It's hard to see how they could be the 'source' of our sensory experiences unless they also 'cause' those experiences.

However, without time, there is no cause and effect, no choice and no free will. If causality lies *only* in the 'hand of God' – his Ineffable Divine Plan – this creates problems about free will: 'predestination'. Though Kant tried, he couldn't escape his awful childhood education in Pietism, a fundamentalist Evangelical sect which slithered towards 'brutal predestination': as in Calvinism – 'the Elect' who reach Heaven are chosen before they're born.

Kant helps us understand the status of a concept, with which we can work out what's real. How do we know the science of climate change is real? We look at cause and effect. What happens when we look at dreams or other timeless inner-world experiences? Are they real? Confused? That's an honest response, as the boundary between real and unreal is *fractal*. Imagine a cloud, or a coast: where is the edge? We imagine we've seen it, but the closer we look, the more the pattern repeats. Symbols also lack edges; in part, because they are timeless. Any symbol on an alethiometer has endless interpretations.

Freud reflected on Kant's *Critique of Pure Reason*. The French philosopher Paul Ricoeur (1970) points out the psychoanalytic project is full of cause-and-effect thinking, as in the much-parodied interpretation – 'it's your mother'. Comparing Freud to Marx and Nietzsche, Ricoeur suggests the three are a 'trio of suspicion', believing our unconscious seeks to conceal or disguise experience: dreams and symbols *conceal* rather than *reveal* truth. Symbols are *only* distorted reflections of basic (sexual) desires. It is as if Freud assumes our unconscious lies. This is not probable. More likely, our unconscious strives to tell the truth. Where will we turn to find out how to face into climate change? To our unconscious, particularly our collective unconscious, which navigated the Ice Ages and the Great Thaws. The idea of the collective unconscious is another Bergson and Jung had in common. Let's see how their ideas came together.

Jung and Bergson

How did Jung happen on Bergson's ideas? I'm indebted to research by Pete Gunter from the University of Texas (1982) for what follows. Jung read Bergson's *Time and Free Will* in 1907 – Bergson was already famous – and discussed it with friends in Zurich. When he distanced himself from Freud, around 1913, a big argument was over the meaning of 'libido'. Did this simply mean sexual energy? Could it mean *elan vital* (vital energy)? This, Bergson suggested, was the differentiating substance between spiritual and material realities. Maybe 'Imprisoned Splendour' is a synonym for *elan vital*.

It's well known that Jung could not 'reduce' dreams to sexual fantasies, or hold religious experience was '*only*' an elaborate deception by a perfidious unconscious. By 1913, he was working on *The Red Book*, internally torn apart by the conflict with Freud. He symbolised it as between 'the Spirit of the Times' (scientific time, and materialism), and 'the Spirit of the Depths' (inner time, and spirituality) (see Jung, C.G., 2009, pp. 229–31). These concepts derive from Bergson. Jung lived out and lived through his own materialist/spiritualist dilemma. The inner world he found was a vivid landscape peopled with mythological figures: Salome, Elijah, Philemon and the dying god, Izdubar. *The Red Book* is unfinished. It gives us a method as useful as free association, called active imagination (see Chapter 8). Both Jung and Bergson use symbolic thinking as a meaning-making bridge between external and internal space-time. What might happen if instead of 'fighting climate change', we wondered what it symbolises?

Symbolic thinking

For Bergson and Jung, symbols are ways to reveal, not conceal, truth. Dreams talk to us in symbols, in timeless language based on archetypes. An archetype can be thought of as a shared pattern for psychological behaviour: for example, the patterns 'mother' and 'father' are shared by all mammals. Forming and using symbols requires intuition, by which Bergson means 'going inside the skin'. He uses the word 'sympathy' to mean 'being inside the object' (Lapoujade, 2018, pp. 39–40). The same happens in empathy, in an analyst's 'counter transference' – sharing sensations and feelings with our patients, which resonate with past experience. We leave 'clock time' and enter duration (*la durée*) – a fifty-minute hour can 'last' fifty seconds or fifty years. Time is a 'both/and' experience: reality is a 'both/ and' experience, too. Reality testing happens in the immediate present – which William James called 'the specious present' – it takes time for neural impulses to be transmitted.

It is 'the prototype of all conceived times . . . the short duration of which we are immediately and incessantly sensible' (James, 1893, p. 609).

Paranormal events and symbolic events occur on or in the borders between the material and the spiritual, sharing something of both. They are liminal. Materialist science tends to ridicule this: if we can't measure it, then it's not real. How do you measure a feeling? With a micrometer? Bergson's solution to the spiritual/material problem is that both are 'real', though it depends how you look at it, and what you are looking at it for. We live in a liminal time; perhaps answers may come from liminal experiences: dreams, active imagination and symbols.

Jung and his friend, the pioneer of quantum physics Wolfgang Pauli, developed thinking around the spiritual/material problem in what's called the 'Pauli/ Jung conjecture' (Atmanspacher and Fuchs, 2014, pp. 1–7). This offers a pluralistic view, a 'both/and'. An 'Imprisoned Splendour' is the 'both/and' construction. This allows more than one thing to be true at the same time. I think this is part of the natural philosophy of symbols – as a symbol, like the position, mass and charge of an electron, can never be completely defined. Symbols contain an 'X', where 'X' means an unknowable; not something we could measure (like the weather on Neptune), but something we can never measure – like a feeling, or 'the Spiritual'. Using 'both/and' defines pluralism: a natural antidote to fundamentalism which is an unwavering attachment to an irrational set of beliefs -which psychiatrists call a delusion. As Anthony Giddens, then Director of the London School of Economics said to *The New Statesman* magazine (Giddens, 1997; italics in original):

Q: What do you consider the greatest threat at present to individual freedom and liberty?

A: The rise of fundamentalism of all kinds. Contrary to received wisdom of the moment, I believe we should oppose all forms of moral absolutism. The simplest way to define fundamentalism is as a refusal of dialogue – the assertion that *only* one way of life is authentic or valid. Dialogue is the very condition of a successful pluralistic order.

The Pauli/Jung conjecture

The Pauli/Jung conjecture is an example of scientific pluralism: an applied 'both/and'. Austrian physicist Wolfgang Pauli (1900–1958) was, like Bergson, a youthful genius, publishing a paper on Einstein's theory of relativity in 1918, when he finished high school. He met Jung in 1930, after a difficult divorce and the death of his mother. Jung referred him to a colleague, kept his friendship, wrote about his dreams in 'Psychology and Alchemy' (CW vol 12, discussed further in the Jung Pauli letters: Jung and Pauli, 2015 *passim*).

They collaborated, among other things, on aspects of the spirit/matter problem and synchronicity, which I'll briefly discuss later. Their conjecture supposes that mind and matter are quantal (exist in minimal quantities). As light can be both a particle and a wave, so 'substance' can have spiritual and material properties: depending on who is observing, and for what purpose. Rather, as Raynor Johnson describes (1953, pp. 190–217), ghosts tend to vanish when you go looking for them : the intention affects the observation. They suggest spirit and matter are complementary and mutually exclusive, yet both needed to describe the natural world – to do natural philosophy, to understand climate change.

Formally stated, the 'Pauli/Jung conjecture' or 'double aspect theory' says different aspects of perception show a complementarity in a quantum physical sense. As the Swiss analytical psychologist Harald Atmanspacher (2012) says, 'Two descriptions are complementary if they mutually exclude each other yet are both necessary to describe a situation exhaustively'. This amplifies Bergson. Another example of dual aspect theory is synchronicity, which Pauli helped Jung clarify.

Synchronicity

One of the strange things about Pauli was his legendary 'paranormal' ability to cause practical physics experiments to break. Apparatus would mysteriously fall apart, go wrong or smash:

> in Professor J. Franck's laboratory in Göttingen . . . without apparent cause, a complicated apparatus for the study of atomic phenomena collapsed. Franck wrote humorously about this to Pauli at his Zürich address and, after some delay, received an answer in an envelope with a Danish stamp. Pauli wrote that he had gone to visit Bohr and at the time of the mishap in Franck's laboratory his train was stopped for a few minutes at the Göttingen railroad station.
>
> (Gamow, 1966, p. 64)

A synchronicity is always a memorable felt experience; an event is synchronicity when a meaningful coincidence occurs within a close time, without causal connection. Jung suggested that there is an 'acausal connecting principle'. His idea originated in the 1920s (CW 8, para. 816–62, and Main (1997) *passim*). Meaningful coincidences are common in analysis; we see patterns of meaning in symbols repeat across a lifetime and over generations. Cultures have synchronicities, too: like realising 'planting trees helps prevent climate change'. One person realises, then everyone has the same idea. Intuition flows naturally through the collective unconscious without advertising or marketing. Suddenly, 'everyone knows'.

It may look like 'the tree-planting revelation' is *only* a 'but we always knew that' – perhaps cultural synchronicities take longer to notice? Cultures are like holograms: each person carries the whole picture. London Jungian Louis Zinkin (1987) describes the brain as a hologram: the capacity to make meaning, indeed meaning itself, is distributed across the whole brain, rather than localised in one part.

Perhaps synchronicity is a 'cross over' state, in which internal time is experienced as external – we projectively identify a need for meaning into the world, then find it, as if the meaning was implicit. David Bohm, Professor of Theoretical Physics at Birkbeck College London, also proposed the value of using two different frames of reference to describe the paradoxical behaviour of subatomic particles. Like Bergson's ideas about the difference between the quantifiable and the qualifiable, between clock time and felt time, and in the Pauli/Jung conjecture, he suggests that the difference between implicate and explicate order relates to differences in behaviour as scale changes. At a subatomic scale, space and time are not good describers of position or interaction. Space, time and causality are enfolded within reality, creating the ground of being (Bohm, 1980, p xv). The 'material world' we live in and experience is 'explicate', folded out from this.

This enfolding of meaning is described by the American psychologist Lawrence Le Shan in *The Medium, the Mystic and the Physicist* (1974, *passim*) who gives many examples of synchronicity. Jung's idea is the projection of unconscious content onto and into the material world is not surprising. He imagined this happened in the paranormal – an intuitive, future dimension. As Le Shan explains (ibid., pp. 62–79), 'trusting the science' no longer must mean 'abandoning a spiritual perspective' – it means regaining a sense of wonder, remembering science includes the imaginal and the intuitive. The Pauli/Jung conjecture simply gives words to this. Synchronicity is a principle which has explanatory power for Jung's concepts of archetypes and the collective unconscious.

I'm not suggesting awareness of synchronicities or 'the spiritual' will 'solve' climate change. I am suggesting our capacity to make new meanings from old symbols is far greater than we imagine.

Time and climate change

'But it is too late. We've run out of time. We're doomed!' No, not really. Time is granular at the smallest level – and time is both a felt experience and a scientific fact. It is a descriptive name for a frontier between material and spiritual experience. We can't talk about 'the timeless present' or 'the eternal now' if we don't have a concept of time in the first place. Time is the space where choice occurs. We don't have to get stuck in a loop in time by making the same mistakes all over again. I'm not suggesting we will be saved by 'the Sprits in the Sky' – a transcendent defence. We need to engage with 'things as they really are'.

Analysts see these loops of trapped meaning occurring in 'projective identifications', one of the primitive defences of the Self. All of them appear when a person, culture or community are under serious threat. Projection means seeing in other people things which belong to us, usually negative things we'd like to deny (like our greed, hatred and delusions). Jungian analysts call this archetype 'the Shadow' – 'everything we would not wish to be' (CW 9 ii, para. 13–9). It can be as simple as feeling that 'it's all their fault' or 'it's all the fault of *(fill in the blank)*' – say, '*greedy capitalism*' (but where is your pension fund investing?) The Shadow also contains our potential, our unlived lives: it's our psychological 'Pension', something we unconsciously draw down as required. So, 'solutions' to climate change exist in the Shadow, and 'out of time' – in inner time, graspable by intuition.

Projective identification differs from projection. This occurs if an unconscious 'particle' leaves one person (or group) and appears in the unconscious of another, producing an immediate, time-free response – before the other's conscious engages: for example, unspoken hostility in body language often provokes a physical retaliation. Unspoken meanings are shared, and there are causal, yet unconscious, connections. In synchronicity, there is a conscious, non-causal connection: the opposite. Much of the 'denial discourse' works by projective identification. A 'mighty leader' is found, promising safety, with ideas which are simple, uncomplicated – and wrong. Truthful answers are complex and need 'a library' – like the one in which I met George in my dream. 'Truth' is an 'Imprisoned Splendour'.

Conclusion

I looked at the concept of time, linking Bergson's idea of quantifiable (scientific, material time) and non-quantifiable (spiritual, inner time), to Jung's vision of 'the Spirit of the Times' and 'the Spirit of the Depths'. I introduced the Pauli/Jung conjecture, to marry the long-standing split between materialist and spiritual views: this is one reason I joined the Scientific and Medical Network as a youth. Analysis helped me learn how to use intuition – to have a better sense of when it was right or not by learning how to form and use symbols. The time-free symbolic language of the unconscious lets us imagine new possible futures (see Chapter 12 in this volume).

The conjecture also explains synchronicity and parapsychological experiences, and why these don't 'behave' under laboratory conditions. They may be 'loops in the fabric of time' – poetic experiences which can't be measured with a micrometer – but could be measured with an alethiometer. In synchronicity, meaning emerges because in such rare moments 'scientific' time and 'inner' time overlap, becoming symbolic time – an internal event experienced externally.

Managing change by inducing guilt is never going to work (See Chapter 4 of this volume). If guilt and shame are projected, the result will be resistance. To adapt to climate change, we need a dual temporal vision. We could see *both* scientific time (when the ice caps will melt) *and* inner time (how long we feel distress when they melt) as valid. We don't have to do the same thing repeatedly imagining we will get a different result, which is how a psychiatrist defines a neurosis. I began with a dream; I've amplified the dream. It surprised me, because this was not what I imagined writing.

It changed my perception. I had thought climate change was to do with greed. Now, I think it also has to do with a natural perceptual limit. We can't meet future people. We can't usually know forebears or descendants more than three generations away. I might meet my great grandchild, but I can never meet their grandchild, a future person. The 'value' argument about moral obligations to future people implies we can leave them things of value, rather than burden them with 'rights'. I'd leave them my analytic toolbox, which I value. There's an alethiometer in it.

Other analytic tools for imagining possible futures include free association, reverie, creative listening, active imagination and dream work. If we learn to form and use symbols, then we can play with meaning. There already are unconscious patterns about facing catastrophe and adapting, many are in the Shadow, in liminal experiences (see Chapter 9 of this volume). Patterns exist in our collective unconscious to cope with catastrophic change. We have used them before. They are time-free experiences, open to everyone. They are 'an Imprisoned Splendour'.

We need to listen to the science, which is 'both/and' 'material /spiritual'. The transcendent is not escapism; it is innate to our shared experience of inner time. We are not running out of time. Inner time is as big as the collective unconscious – at present, this has a size of about seven billion lifetimes, which is about fourteen billion hours of dreaming every single night the planet turns. Nor are we running out of ideas. Time is one defining characteristic of causality. Dreamed, imagined and intuited actions also have consequences. Once I imagined editing a book about depth psychology and climate change, a tiny quantum toward a change in dominant cultural attitudes. Each of us can make a quantal difference. This means giving our time, external as well as internal. Now, let's go and plant some trees.

Acknowledgements

My sincere thanks to Grant Clifford, Grace McLean, Dr. Carola Mathers, Dr. David Mathers, Mary Jayne Rust and Andy White for their helpful comments on earlier drafts of this chapter.

References

Atmanspacher, H. (2012). Dual-aspect monism a la Pauli and Jung. *Journal of Consciousness Studies*, 19(9–10), pp. 96–120 (25).

Atmanspacher, H. and Fuchs, C. (2014). *The Pauli – Jung conjecture and its impact today*. Exeter: Imprint Academic.

Bergson, H. (1912/reprint by Forgotten Books, 2018). *Time and free will: An essay on the immediate data of consciousness*. London: George Allen.

Bohm, D. (1980). *Wholeness and the implicate order*. London: Routledge.

Deleuze, G. (2004). Bergson's conception of difference. In: *Desert Islands, and other texts*. Paris: Semiotexte.

Ellison, A. (1988). *The reality of the paranormal*. London: Harrap.

Gamow, G. (1966). *Thirty years that shook physics – The story of Quantum theory*. New York: Doubleday & Co.

Giddens, A. (1997). *Brasher, Steven 'Influences'*. London: The New Statesman, pp. 32, 126, 4319.

Guerlac, S. (2006). *Thinking in time: An introduction to the work of Henri Bergson*. Ithaca, NY: Cornell University Press.

Gunter, P.A.Y. (1982). Bergson and Jung. *Journal of the History of Ideas*, 43(4), pp. 635–652, Pennsylvania, PA: University of Pennsylvania Press. Available at: www.jstor.org/stable/2709347.

James, W. (1893). *The principles of psychology*. New York: H. Holt and Company.

———. (1909/2015). *A pluralistic universe: The Hibbert lectures, at Manchester University*. London: reprint by Erik Publishing: Amazon, UK.

Johnson, R.C. (1953). *The imprisoned splendour*. London: Hodder and Stoughton.

Jung, C.G. (1953–77). *Except where indicated, references are by volume and paragraph number to the collected works of C. G. Jung*. 20 vol., ed. by H. Read, M. Fordham, and G. Adler, trans. R.F.C. Hull. London: Routledge and Princeton: Princeton University Press.

———. (2009). *The red book*, ed. by S. Shamdasani. London: W.W. Norton & Company.

Jung, C.G. and Pauli, W. (2015). *Atom and archetype: The Pauli – Jung letters, vol.1*. Princeton, NJ: Princeton University Press.

Lapoujade, D. (2018). *Powers of time, versions of Bergson*. Minneapolis MN: Minneapolis University Press, Univocal.

Lawlor, L. and Leonard, V. (2016). Henri Bergson. In: *The Stanford encyclopaedia of philosophy*. Stanford, CA: Stanford University. Available at: https://plato.stanford.edu/archives/sum2016/entries/bergson/.

Le Shan, L. (1974). *The medium, the mystic and the physicist*. London: Turnstone Press.

Main, R. (1997). *Jung on synchronicity and the paranormal*. London: Routledge.

Mathers, D. (2001). *An introduction to meaning and purpose in analytical psychology*. London: Routledge.

Nagy, M. (1991). *Philosophical issues in the psychology of C.G. Jung*. Albany, NY: State University of New York Press.

Pullman, P. (1995). *Northern lights*. London: Scholastic.

Ricoeur, P. (1970). *Freud and philosophy: An essay on interpretation*. London: Yale University Press.

Rovelli, C. (2018). *The order of time*. London: Allen Lane.

Russell, B. (1912). The philosophy of Bergson. *The Monist*, 22, pp. 321–347.

Skinner, S., Prinkle, R., Hedesan, G. and Goodwin, J. (2019). *Splendor Solis; The world's most famous alchemical manuscript*. London: Watkins.

Stein, L. (1957). What is a symbol supposed to be? *Journal of Analytical Psychology*, 2(1), pp. 73–84.

Weiner, Y. (1996). Chronos and Kairos two dimensions of time in the psychotherapeutic process. *Journal of the British Association of Psychotherapists*, 30(1), pp. 65–85.

Windle, P. (1995). The ecology of grief. In: T. Roszak, M.E. Gomes, and A.D. Kanner, eds., *Ecopsychology*. Berkeley, CA: Counterpoint.

Wright, C. (2018). *Obligations to future people: Thesis for B.Phil.* Oxford: Oxford University Press.

Zinkin, L. (1987). The hologram as a model for analytical psychology. *Journal of Analytical Psychology*, 32, part 1, pp. 1–21.

Website

The Scientific and Medical Network: https://explore.scimednet.org/.

Chapter 12

21st-century unconscious

Altered states, oracles and intelligences

Joe Cambray

Introduction

A contemporary depth psychological approach to ecology envisions a non-local, distributed psyche (see Figure 12.1). The human aspect of the psyche, including mind, is necessarily embedded in the larger world of nature. Such an expanded vision of psyche and mind requires some revision of the nature of 'the unconscious'. This will have multiple components, including the biological, the cultural and the archetypal, as well as the personal.

In stepping beyond a one-person, intrapsychic view of the mind, Jung in 'The Psychology of the Transference' (CW 16, paras. 353–539) described what can be recognized as a field approach. Psyche is encountered in an interactive field between the therapeutic partners (analyst and analysand). Extending this to an engagement with the world (human and natural) leads to a view of the psyche distributed in a network of relations with people and objects in the surrounding environment. Combining this with Jung's structural vision of the unconscious described in layers of increasingly broader inclusiveness of others (family, tribe, community, culture, humanity) which also is layered historically, we arrive at a dynamic psyche distributed in the field of interactions according to archetypal patterns. By including the psychoid aspects of the archetypes which when activated can correlate with synchronistic experiences, the non-local dimension of the psyche (not constrained to single locations in space or time) is more fully revealed. From its inception, the unconscious has been a mercurial notion, never to be definitively codified in a final form, shape-shifting in its expressions through time and space. This includes its theoreticians' perspectives, involving implicit cultural and collective dimensions of the concept. There are some recent new approaches to the mind which help us seek a contemporary language, in particular from complexity studies and the idea of emergence. The latter is a property of complex adaptive systems (CAS), those in which agents interacting near the edge of order and chaos are capable of spontaneous, self-organizing transformations producing new holistic forms which cannot be reduced to a sum of the properties of the components (Figure 12.1).

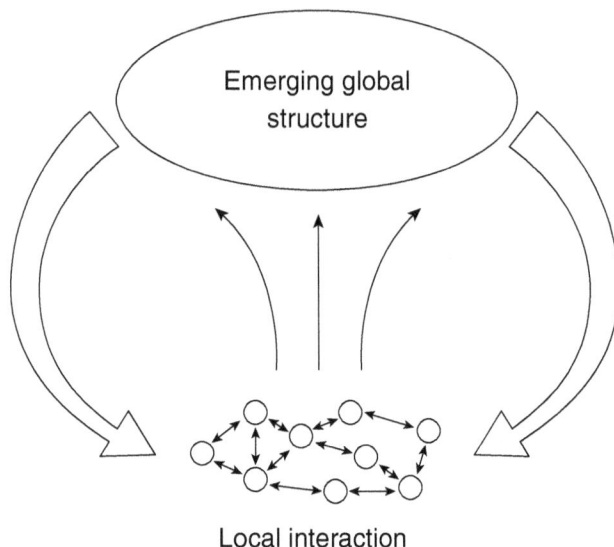

Figure 12.1 The emergence field

Source: Ratter (2013) (available free at 'ResearchGate'; https://www.researchgate.net/
publication/259475046_Complexity_and_Emergence_key_concepts_in_non-linear_
dynamic_systems)

In graphic representations, feedback loops between agents and emergent forms
are usually included. These loops indicate emergent forms impacting the engagement among agents, which in turn modify the emergent form in an interactive,
iterative manner.

The study of emergence transcends algorithmic approaches, and the dynamics
are highly non-linear and cannot be described with equations, only approximated.
The conditions which produce a hurricane or a catastrophic forest fire can be qualitatively described but cannot be accurately predicted through the use of mathematical models – as in the spread of forest fires – non-reducible, holistic forms
are engendered through the interactive engagements among the agents. Emergent
phenomena have been discovered at all levels of scale, from the subatomic to the
clustering of galaxies and beyond. These phenomena are prevalent in biological systems, e.g., the human microbiome with cultural differences/diversity, and
include human behaviors, from traffic jams to stock trading (Taleb, 2010).

These can be stable for long periods (galaxies), or brief, dissolving after serv-
ing their function (e.g., murmurations of starlings at dusk before settling into a
resting place for the night). In philosophy of mind and consciousness studies, the
'hard question' of the link between brain and mind, while unresolved, has been
approached from an emergentist perspective (mind emerging out of the body,
brain, environmental interactive field, which suggests intrinsically non-local,

ecological qualities of the psyche). Culture, then, is at least partially emergent from the interactions of many psyches. In the case of the analytic encounter, the agents are the dyadic partners with their conscious and unconscious contributions, personal and collective, an interactive field in which the analytic dyad resides.

Applying the study of emergence to the analytic situation produced interesting results. In psychoanalysis, one area has been the work of the Boston Change Process Study Group, spearheaded by infant researcher and psychoanalyst Daniel Stern (Stern et al., 1998). In researching the analytic situation, they came to challenge the sole primacy of interpretation and looked at the impact of the interactive relational exchange in transformational analytic encounters. The trajectory of change in psychotherapy they articulated was based on an emergent process model requiring skillful management of intensifying affective moments – navigating from talking about things to entering into discussions framed in the present, then on to 'now' (affectively charged) moments about the present – leading to potentially mutative 'moments of meeting', in which negatively tinged responses are anticipated but instead a new relational pattern emerges drawing upon empathic understanding with relatedness, offering a mutative alternative to the expected pattern.

A consistent series of such 'meetings' can have a powerful therapeutic effect (Stern, 2004). More broadly, the interactive field can and has been (re)conceived along similar lines, expanding the notion of the unconscious into this field, beyond just the intrapsychic, to the interpersonal including non-local dimensions in congruence with Jung's model stated previously. Paralleling these and related developments has been the reconsideration of key Jungian concepts such as archetypes, individuation, synchronicity and transference-countertransference dynamics (Knox, 2003; Hogenson, 2005, 2009, 2019; Cambray, 2002, 2009, 2011; Martin-Vallas, 2008; Merchant, 2019; Coleman, 2016, among others) along the lines of CAS with emergent properties. These approaches alter views on the nature of the unconscious, starting with applications of complexity studies to states of consciousness, which I now discuss.

States of consciousness

Depth psychological approaches necessarily, deliberately engage expressions of unconscious processes and thus require practitioners to alter their ordinary (egoic) states of consciousness to more fully access this material. Classic analytic techniques were designed to assist in this, starting with Freud's striving for free association. To remain an inquisitive interlocutor in this psychic space, the analyst needs to defocus consciousness to some degree and allow subtle, altered states of consciousness to provide data about what is emerging, or is being inhibited from emerging. Thus, it takes considerable training to develop and hone oneself as an 'analytic instrument' through judicious application of the communicative dimensions of the transference/countertransference field. There are clear and obvious parallels with techniques developed within various indigenous traditions, sha-manic journeys being just one variant (see Chapter 9 of this volume).

There have, of course, been important developments, such as in the neo-Kleinian and Bionian schools, whereby the uses of the analyst's reveries (accessing another set of altered state homologous to those found in mother-infant dyads) have become a source of valuable analytic information (Ogden, 1997 *passim*). Similarly, extended explorations of dyadic field phenomena (Ferro and Basile, 2009) point to an expansion of consciousness beyond the individual (Tronick, 2005) into shared, co-constructed states. New York psychoanalyst Jay Frankel has written about play (itself often an altered, reverie state) in a Winnicottian way as 'the deep structure of the state of consciousness that drives the analytic pro-cess' (2011). He traces this back to Sandor Ferenczi, who saw neurotic patients as spontaneously regressing into (quasi)-hypnotic states, which activate transfer-ential imaginings.

Within the Jungian paradigm, training includes exploration of altered states through careful work with dreams, active imagination (beginning with Jung's experiments captured in *The Black Books*, then *The Red Book,* Jung, 2009), and examination of interactional fields (for example, Martin-Vallas' notion of the transferential chimera, *vide supra*) including the role of meaningful coincidences (synchronicities) in the analytic process (see Chapter 11 of this volume). Each of these approaches again involves drawing upon a variety of altered states. These types of activation of unconscious processes tend to engender experiences which give evidence of non-local dimensions of the psyche. Previously such experi-ences were well beyond the sciences of the early 20th century, when analytic schools were forming and codifying their models, so the phenomena tended to be marginalized or dismissed. This was a time when explorations of the margins of consciousness were linked, often dismissively with occultism and the paranormal. Nevertheless, societies for psychical research (from 1882 to the present) have had numerous prominent figures on both sides of the Atlantic actively involved. A short list of presidents of the Society for Psychical Research includes William James, Fredrick Myers, Sir William Crookes, Andrew Lang, Henri Bergson, Gil-bert Murray, Hans Driesch, E.R. Dodds and J.B. Rhine (see www.spr.ac.uk/about/past-presidents); it is still in operation.

Complexity studies opened a pathway to reconsider the non-local aspects of psyche as holistic features deserving more careful attention. For this to occur, we will likely need to expand our view of the range of unconscious phenomena being considered, including but going well beyond the dynamically repressed. Similarly, these studies are linked to ecological systems and so when considering climate change, especially the psychological impact which can be described as climate trauma (Woodbury, 2019), suggest an additional expansion of the uncon-scious. In doing so, our attitudes could be valuably augmented by inclusion of more dynamic, non-literal understandings of altered states of consciousness.

There has, of course, recently been a strong resurgence of interest in altered states in the United States, especially those induced by (bio)chemical means (psychotropic plants and their active ingredients) as well as through meditation, sensory deprivation and other techniques. There also is a revival of broad interest in the psychotherapeutic potential of these altered states, especially in the treatment of situations not historically amenable to analysis: substance abuse, addictions, refractory depressions and the existential dread of dying. Michael Pollan's recent book *How to Change Your Mind* (2018) has become a best seller, with its synthesis of history and research on the topic of psychedelics.

Analogous to what Jay Frankel wrote about 'play', the current research on altered states does point to heightened sensitivities, affective lability, reactivity to stimuli, permeability of boundaries and a tendency towards 'regressed' states. However, in addition, there seems to be a noetic quality, a keen intuitive capacity to respond to knowledge not easily articulated and often out of direct conscious awareness. These phenomena are radical extensions of the realm of the 'unthought known' (Bollas, 1987 *passim*); further, they often have a symbolic, overdetermined quality, so perhaps are better reframed as states with a surplus of knowledge (Struck, 2016) or meaning. William James defined the noetic quality in mystical experiences as:

> states of insight into depths of truth unplumbed by the discursive intellect. They are illuminations, revelations, full of significance and importance, all inarticulate though they remain; and as a rule they carry with them a curious sense of authority.
>
> (1902/2011, p. 312)

This brings us to two research areas which may be of benefit: the study of the use of oracles throughout history, and the exploration of intelligences not associated with a conscious mind. As new adaptive phenomena emerge, facing into problematic futures can be seen as activating abilities humans have always drawn upon: divination, with surplus of meanings being submitted to collective sets of interpretations. Consider how in contemporary culture, artificial intelligence (AI) is becoming much more vast than individual human intelligence – the computers programmed to outplay the world best human Chess or Go players, to name just two of the more obvious. AI can also be understood as adaptive intelligence, but without evidence of sapience or consciousness. To optimists, AI will free us from unwanted work and disease; to pessimists, we will succumb to the power and blind authority of the digital intelligences (Harari, 2017). But in the meantime, how might our views of the mind, and the unconscious be impacted by these developments?

Oracles and intelligences

As our knowledge of the biological world grows, intelligence or cognitive capacities are increasingly recognized throughout nature in the adaptability of organisms to their environments. This often involves problem-solving and optimization of actions in the face of environmental stressors or opportunities. These abilities are not restricted to organisms with brains or even neural systems. Intelligence manifests on the cellular level and tends to rise in complexity as organisms acquire increasing diversity of component cells. At the cellular level, consider slime molds – a single-cell amoeboid organism which can self-organize into a multicellular creature to optimize nutritional intake when opportunities are presented. The gain in complexity manifests in intelligence that can even be engaged in useful ways to humans once its capacities are appreciated.

Hence, in a 2010 publication, a group of scientist and engineers used slime mold to explore the underlying patterning used in building of a rail network: using a map of the city of Tokyo with oat bran placed at the locations of the rail stations, slime mold was introduce at the central main station and allowed to explore the terrain for about 26 hours (seeking out other flakes of oat bran). The network of trails from these efforts bears remarkable first-order similarities to the network model constructed by engineers over about 15 years (Figure 12.2).

Certainly, the slime mold network would have been a valuable first draft – the conclusion of the article, which encouraged finding ways to engage biological

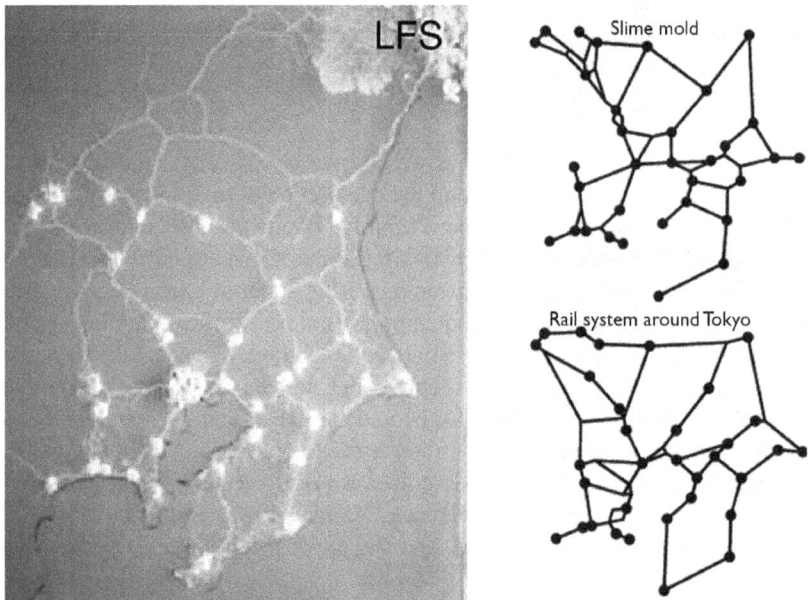

Figure 12.2 The Physarum network

Source: Tero, et al. (2010) (available free at 'ResearchGate'; www.researchgate.net/figure/Comparison-of-the-Physarum-networks-with-the-Tokyo-rail-network-A-In-the-absence-of_fig2_41111573)

adaptive systems in solving certain kinds of problems (Tero et al., 2010). See also Adamatzky et al. (2013), who state:

> the slime mould shows outstanding abilities to adapt its protoplasmic network to varying environmental conditions. The slime mould can solve tasks of computational geometry, image processing, logics and arithmetics when data are represented by configurations of attractants and repellents. We attempt to map behavioral patterns of slime onto the cognitive control vs. schizotypy spectrum phase space and thus interpret slime mould's activity in terms of creativity.

Swarm logic, or the intelligence of collective/communal biological agents self-organized into CAS, is widespread in nature, as in the examples of murmuration mentioned previously. This is true across interacting species as in forests with mycorrhizal networks: fungi synergistically infiltrating tree roots to create vast biochemically communicative networks of trees and fungi. These entities are enormous in total size and weight and have rapid communication capacities through the shared root networks. For example, a beetle infestation damaging to the trees at the periphery of a mycorrhizal network will set off a phyto-chemical/hormonal alarm system distributed by the fungi that allows other tree members of the network to alter their sap biochemistry to ward off the attacks. The work of Suzanne Simard and colleagues has been foundational in this field (Simard et al., 2012), popularized by Peter Wohlleben (*The Hidden Life of Trees*, 2016) and incorporated into a Pulitzer Prize winning work of fiction, Richard Powers' *The Overstory* (2019). Similarly, the human immune system is also capable of biochemical learning, which is why we use vaccines; all of this far from any conscious awareness.

By analogy, are there not psycho-biological intelligences contributing to the unconscious? These can take mutable shapes depending on how we approach the unconscious. Jung commented: 'We know that the mask of the unconscious is not rigid – it reflects the face we turn towards it. Hostility lends it a threatening aspect, friendliness softens its features' (CW 12, para. 29). Through our dreams, we witness figurations of our conflicts and at times attempts at compensation and resolution. Dream imagery often holds relational intelligences including object relations, as we tend to personify in the denizens of our dreams. Active imagination then provides a pathway for engaging these personified intelligences. Developing a symbolic attitude can greatly facilitate useful exploration of dream or reverie presentations from the unconscious. Consider the following dream of a woman in a long analysis with me:

> I was in the home of the "beautiful woman". She had wooden (plank) floors. We were lying on the floor talking when I noticed a brown centipede circling us.

I asked her if I should kill it and she said, "No, it's our friend". We went on talking until I noticed that it was quite close to me and it had something.

She said, "This is wonderful, he's starting to trust you".

He had turned darker and uglier. Then he came between us and crawled on my arm. The "beautiful woman" said I was really lucky, "because look at what he can do". With that he clicked open an ornate box. He held it open for us to look. Inside the box were incredibly beautiful jewels, bright shiny colors. I saw a red jewel, a huge ruby.

The "beautiful woman" said to me, "Whenever I can't find something, I go to him". She loved him.

This proved to be a highly disturbing but pivotal dream in the middle of an analytic process. For the present, I will only comment on the centipede, as I have discussed the dream elsewhere (Cambray, 2011). The centipede symbolized a deep but unconscious ambivalence toward the analyst and the therapy with its attention to the unconscious. The tension between the dream ego and an idealized self-object (beautiful woman) is reflected in their opposing attitudes toward the centipede. In the countertransference, I felt considerable, justified anxiety about the dangers of a psychotic process at this moment; centipedes can have poisonous stings (by analogy, psychosis is as if a mind poisoned). Help in containing the patient and metabolizing the dream was drawn from amplification:

In Tahiti the two indigenous centipedes are regarded as shadows of the medicine gods, and are never disturbed or killed. If one can be induced to crawl over a sick person, that person will surely recover.

(Leach, 1972, p. 206)

The resonance was especially forceful as my first wife, unbeknownst to the patient, was Tahitian on her mother's side. Here we can directly see the intelligence emerging in the therapy via the unconscious field presenting a psychotic activation in an insect form which suggests both grave danger and potential personal development, as well as transferential anxieties, in a surplus of meaning. In retrospect, an especially intriguing statement in the dream is the beautiful woman's pronouncement that 'he (the centipede-analyst) is starting to trust you', not what might have been expected: 'you should learn to trust him'. This dream proved to have an oracular aspect, as we went through a subsequent psychotic episode in the analysis; but unlike her previous episodes, we were able to metabolize much of what was occurring, including working with her rage over a terrible, traumatic history. I was indeed beginning to trust her dreams and she was eventually able to trust the material arising from the unconscious and find a symbolic, therapeutic relationship with it resulting in her remaining free from breakdown for more than 30 years to date.

At times, dreams can have precognitive elements; in my own practice, this has been particularly true in working with trauma survivors. It is as if the normal (linguistic) channels of communication are rendered non-functional because of the fragmentation associated with traumatic injury, along with the intense affects activated, even if not consciously acknowledged, so alternative (more 'archaic') means of communication are found by the unconscious the communicative potential accessible through altered states of consciousness can have great therapeutic value, even though occurring through unanticipated means.

A clinical example I've written about more extensively (2002) is of a case of a woman who survived childhood sexual abuse. During a vacation break, she had a dream which terrified her in which I, the analyst, was not locatable as I was in the Black Forest (she supposed I was in Germany). In fact, the next day I went for an inaugural dive at a site called 'the Black Forest', an exquisite visit to a 'forest' of black coral in the Caribbean Sea: radically different experiences of a shared image unconsciously but profoundly linking us. The impact of this precognitive, synchronistic event greatly enhanced our attunement, symbolically with a subject/object linkage. In general, the Western philosophical distinctions between subjective and objective levels of reality are often blurred or intermixed in these kinds of experiences.

The 120 years of analytic work exploring the unconscious, filled with anomalous phenomena, would suggest a new, more complex paradigm, beyond a reductive, materialistic one, is required if problems of consciousness and mind-body relations are to be seriously pursued using analytic methods. We will need this paradigm as our species adapts to radically new conditions on Earth. The intelligence of the dream offered here reveals what could be termed oracular images – the 'centipede' and the 'black forest' in these cases served a symbolically oracular function. Witnessing many such examples over the years has led me to consider the history of oracles and their function and use. The founders of depth psychology (Freud, Jung and Ferenczi, among others) often drew on the myths, rites and literature of ancient Greek culture in formulating analytic thought, much of which involved oracles (recall Oedipus consulting at Delphi and Freud's use of this). Oedipus sent his brother-in-law, Creon, to the oracle at Delphi to learn how to help the city. Creon returns with a message from the oracle: the plague will end when the murderer of Laius, former king of Thebes, is caught and expelled; the murderer is within the city (see www.sparknotes. com/drama/oedipus/summary/).

Laius, his father, had already been warned by the oracle at Delphi that he should remain childless or else his child would murder him and marry his wife. I would suggest various elements of this are implicit in our analytic attitudes – recall Little Hans, after having been interviewed by a seemingly omniscient Freud who had told him his conflict was predicted before he was born, asked his father: 'So does the Professor talk with God then so that he knows everything ahead of time?' (Strachey, 1955, pp. 42–3). The oracular imagination and its place in the formulations of the unconscious deserves new attention in the light of complexity studies, as well as new understandings of the role and function of oracles in the ancient world.

An oracle is generally a person/figure capable of dialoguing with the gods (archetypes of the unconscious), usually in an altered state and making pronouncement of knowledge from this aspect of contact with the unconscious. The knowledge itself often is ambiguous and needs interpretation; famously throughout history, some dreams have been viewed as having oracular qualities requiring interpretation frequently of a symbolic nature. A few historical examples of dreams with oracular qualities: in the Bible, Joseph famously interpreted the Pharaoh's dream of the seven fat cows followed by seven lean ones coming up out of the Nile, who then ate up the fat ones, as coming from God and foretelling years of feast and famine. The interpretation, which was accepted, allowed for planning that saved Egypt from starvation as well as generating wealth.

Several days before his death, US President Abraham Lincoln had a dream which ended with his asking:

> "Who is dead in the White House?" I demanded of one of the soldiers. "The President" was his answer; "he was killed by an assassin!" Then came a loud burst of grief from the crowd, which awoke me from my dream.
> (see www.history.com/this-day-in-history/
> lincoln-dreams-about-a-presidential-assassination).

And, a third:

> New York lawyer Isaac Frauenthal had a dream before boarding the RMS *Titanic*. "It seemed to me that I was on a big steamship that suddenly crashed into something and began to go down". He had the dream again when on board the Titanic and was alert to the danger when he heard about the iceberg collision. Frauenthal survived the sinking.
> (see https://en.wikipedia.org/wiki/List_of_dreams)

Scholarship on ancient cultures has in recent years begun to look more seriously at the role of divination in the long-term political stabilization of various cultures. Several dimensions of divinatory practices outside prognostic success have been noted in this regard. Thus, consulting the stars, common in many of these cultures:

> demanded incessant reflection upon political and military activity, generating thereby an atmosphere of political vigilance.
>
> The prediction of failure and defeat necessitated the ever-renewed study of the administration, army, and security services, as well as the random verification of the trustworthiness of counselors and allies alike. . . . Although society possessed a strictly hierarchical character, it was impossible for even persons of the highest rank to exempt themselves from scrutiny of their area of responsibility.
> (Maul, 2018, p. 257)

Through these means even those in a counselor's role, those who would not ordinarily be able to oppose the ruler, 'could present their opinion openly without gaining a reputation for being disloyal, ungrateful, insubordinate, or unfaithful, and they could work to ensure that their proposal would be made the object of a new oracular question' (ibid., p. 259). Thus, the use of divination brought levels of increased complexity to governments and social systems. When optimized, this complexity helped to sustain the structures themselves. It was an important aspect of the long-term stability of the societies. A further example is found in the origins of democratic forms of government inaugurated in Athens by Cleisthenes, during the crisis of the wars with Sparta (see Cambray, 2009, pp. 88–92) facilitated by support from Delphi – Aristotle on the Athenian Constitution part 21: 'The names given to the tribes were the ten which the Pythia appointed out of the hundred selected national heroes' (Aristotle, n.d.).

Divination was generally oriented toward pattern recognition with interpretations of those patterns based on accumulated, historical observations. Even though the interpretive content was not always accurate, and was obscure or ambiguous and open to significant misinterpretation, the efforts involved produced a training of observational skills: the search for patterns of increased complexity emerging from interacting with the world that proved to be useful to identify. Even if some identified patterns were projections, they were revelatory of psyche, and at times, psyche and world overlap in significant ways. Correct interpretations of patterns were often life-and-death matters and so were given much importance. Hence, those involved tended to develop powerful intuitive functioning in response. Of particular interest here was the use of altered states in making oracular pronouncements. Perhaps a useful reframe would be to see oracles as accessing some of the intelligences in the unconscious and presenting them to conscious awareness for our contemplation and reflective considerations.

The most famous oracular center in the ancient Greek world was at Delphi, where the Pythia offered consultations; it was considered one of the most sacred sites in ancient Greece from about 1400 BCE until 400 CE. There is now ample proof the tripod she sat on when making her prophecies was directly atop a geological fault which emitted hydrocarbon gases, including ethylene (Johnston, 2008, pp. 48–50). This was used in surgery as an anesthetic until the early 1970s. Its sweet odor and psychoactive properties, which can cause hallucinations even in low doses, match detailed accounts of witnesses of the consultations with the Pythia, such as Plutarch, who was both an historian and a priest of Delphi. The reinstatement of the validity of the actual site may reflect some of the changes occurring in culture as we move into the 21st century with new, more complex ways of seeing and understanding anomalous phenomena generally.

The Pythia was imagined to be influenced by the god/s, in particular Apollo, through the 'pneuma' she inhaled. The aery realm was seen as the domain of spirits, where the soul might converse with the daimones and gods. According to Sarah Iles Johnson, American Professor of Classics,

> Democritus explained divinatory dreams through a theory of eidola. He saw
> the things of the world sloughing off eidola (a spirit-image or double), which

can penetrate the soul of sleeping persons; this included eidola from another person's thoughts or feeling, which leads to a form of precognition.

(ibid., p. 15)

This even seems to have a seasonal or climatic component, as Johnston notes, according to Cicero: 'In autumn, when the air is rougher than usual, these eidola don't travel very well and our dreams are therefore rather faded and ineffectual' (ibid.), perhaps a folkloric description of a non-local psychic field (distributed in the aery/imaginal realm) in which reception is impaired by an agitated medium. Here calm 'air' (a tranquil mental state in the face of fluid imagery) is conducive to perceiving this field whereas 'autumnal' conditions (disruptive transitions, toward the dark and cold) disturbs and inhibit reception of the non-local imagery. This further hints at an emergent, ecological vision of the psyche in concert with contemporary complexity studies applied to archetypes and the psyche more generally.

From the numerous accounts and the fame of the Pythia (ancient peoples throughout the Mediterranean basin journeyed to consult her at Delphi), there is ample reason to accept that she had remarkable prognostic abilities in an altered state of consciousness. For the Greeks the term for divination was *mantike*, akin to mania, meaning an inspired or divine madness. Certainly, depth psychologists have recognized the deeper truths in her pronouncements, such as to Laius and Oedipus. If we shift the focus from the literal veracity of prophecy to psychological and symbolic understanding of contents arising from the unconscious, we find clear parallels with the role of the analyst in listening to the unconscious and speaking out of the material, the 'intelligences' contacted in the process. Altered states can therefore be useful for knowledge rather than simply prophecy, yet analysts often 'foresee' psychological issues emerging from the unconscious which, if not addressed, can lead to difficult at times tragic consequences.

In a related formulation, I discussed theoretical biologist Stuart Kauffman's idea of the 'adjacent possible' (Cambray, 2019), which could be applied here to anticipatory foreknowledge. The 'adjacent possible' is the virtual network of possibilities one step away from where we (individual or collectively) are aware. For the individual, a step into the 'adjacent possible' creates novelty; for a society, it is the path to innovation as is needed now. Dreams often tap into this, as when we dream of finding a new room in a building (often our own home) which we are familiar with – the new space if entered offers pathways into the novelty. Opening one door often leads into a sequence of new rooms, expanding consciousness through navigating the adjacent possible.

Returning to the environment of Delphi, even the physical situation of the site, as at many other places where prophecy occurred, is worth consideration as psychologically symbolic. Prophetic centers were often found in clefts in the earth, or in caves, in twilight realms, where noxious vapors arise and intoxicate. This can be read psychologically as activity occurring in proximity to an activated unconscious, with its seemingly haunting spirits and uncanny pronouncements. These are often hard to decipher and understand, requiring a skillful, trained interpreter who nevertheless is working in a realm of surplus meanings, frequently with a sense of mild disorientation while attempting to select the most relevant threads to present in a tactful manner; the parallels to analytic work and creativity are evident.

The 'future foreseen' could be understood as the intuitive trajectory of activated psychological forces at play in the personality of the 'seeker', as this impacts the 'interpreter'. This is paralleled in contemporary views of the therapeutic field. Again, non-local, distributed aspects of a shared state of consciousness seem a key component in this model, which opens to the environment and a more ecological view of the psyche itself. Such a stance implies a change in worldview, exploring the unconscious mind with a hermeneutic science, while drawing upon logical, positivistic scientific information (as from the neurosciences) further informed by systemic, holistic hermeneutics. In the conjunction, we could bridge the gaps opened up by the enlightenment approaches. The depth psychological attitude derived from this movement began by Freud, was expanded upon by Jung and others, and could now be taken another step further.

In a similar manner, there was a profoundly ancient oracle tradition in China, starting with 'oracle bones', the shoulder blades of oxen or plastrons of turtles (the flat, underside of the turtle's shell) used in the Shang Dynasty of China (c. 1600–1046 BCE) for divination. The practice evolved in an especially powerful way during the Zhou dynasty (1046–256 BCE) with the development of the *I Ching*. Refined and employed by some of the best philosophical minds in China throughout the centuries, the *Book of Changes* meaningfully entered the analytic/ depth psychology world through the translation into German by Richard Wilhelm in 1923, for which Jung eventually wrote a foreword in 1949.

Since his receipt of another of Wilhelm's translations, the Taoist alchemical *Secret of the Golden Flower* in 1928, Jung had grown increasingly interested in the complementary nature of Chinese philosophy to Western science. He realized the incompleteness of the purely causal Western approach to understanding reality. Valuable insights gained by attending to the quality of a given moment, paying heed to the coincidences occurring in time, are fundamental to the methodology of the *I Ching*. Jung commented:

> The moment under actual observation appears to the ancient Chinese view more of a chance hit than a clearly defined result of concurring causal chain processes. The matter of interest seems to be the configuration formed by chance events in the moment of observation, and not at all the hypothetical reasons that seemingly account for the coincidence.
>
> (Jung CW 11, para. 969)

Here, Jung is opening our analytic attitude to awareness of meaningful coincidences, especially when analyzing the field in terms of the whole situation in a given moment. This can be read as a psycho-ecological formulation of mind;

applicable as in his famous clinical demonstration of the scarab beetle dream synchronicity story (CW 8, para. 843). The question of the intelligence(s) in the *I Ching* oracles arises for Jung in terms of explaining to uninitiated westerners his experience in using it through personification of the book itself:

> according to the old tradition, it is "spiritual agencies", acting in a mysterious way, that make the yarrow stalks give a meaningful answer. These powers form, as it were, the living soul of the book. As the latter is thus a sort of animated being, the tradition assumes that one can put questions to the *I Ching* and expect to receive intelligent answers.
>
> (ibid., para. 975)

Here we can observe Jung's articulation of what I call the 'intelligences of the moment'. Discerning the adaptive qualities implicit in the gestalt of a moment reveals the potential to access a profound and guiding intelligence which can be constellated in the unconscious as a cosmic reflection of the moment. Perhaps we have here a method to glimpse the intelligence operating in moments of complexity (an extension of moments of meeting which I wrote about in 2011). If accurate and resonant, this expands the view of the unconscious to include the agencies at play in psycho-ecological fields. I further suggest that this could provide a pathway for the conscious reincorporation of elements of indigenous psychologies back into our reflections on the meaning of 'depth'.

Attitudes toward the unconscious revisited

With a vision of an unconscious animated by intelligences, at times with oracular powers operating outside the ordinary boundaries of time and space, expressing transformative potentials through emergent forms, how do we seek an optimal attitude? In addition to typical analytic and meditative techniques, a suspension of conscious knowing, allowing amazement and surprise to enter (as indicators of emergence) helps open our attitudes to that which does not conform to our preconceptions. Sustaining this attitude is difficult as cognitive interpretive impulses enter rapidly but can be successfully resisted and suspended temporarily, with training and effort. This stance also opens more affective access to traumatized states with the various anomalies to which they give rise.

A further consequence of expanding the states of consciousness to be included in our depth approach is the necessary blurring of subjective and objective distinctions. Cosmologies of deep inner world explorations curiously at times reflect those derived from astrophysical observations and mathematical analysis (see for example Le Shan, 2012). We could also look at the way artists throughout history have produced works with amazing insights into nature that science has taken centuries to replicate – for example, observation of Penrose tiling (named after the cosmologist and mathematician Roger Penrose, who first described this form of aperiodic tiling, reaching symmetry only at infinity) – on 13th-century mosques in Isfahan, as

I described in other publications (Cambray, 2009, 2017). This led me to suggest a psychoid imagination whose productions are accessible by some highly gifted individuals. I coined this term to describe the imaginative capacities of selected artist to represent highly complex patterns found in the natural world, generally before they have been described mathematically or scientifically. Artists who accessed psychoid levels of reality include, for example, Jackson Pollock, who through his drip painting found his way to what subsequently has been recognized as fractal, well before they were articulated by Beniot Mandlebrot, and reflect Pollock's successful search for optimum aesthetic impact based on fractal density on a planar surface (du Sautoy, n.d.). Similarly, Van Gogh's paintings of times of turbulence in his own psychological life mirrored, though not necessarily causally, his ability to capture turbulent phenomena (St. Clair, n.d.). As we have learned, the psyche can move effortlessly between these descriptions, and so our attitude toward what we attend to needs to reflect this fluidity and interpenetration of subjective and objective. Perhaps not collapsing levels of meaning in what is heard during therapy sessions, or even between them, offers us an opportunity to appreciate the fullness of psychic reality and its relationship and expression in external reality. These insights could be productively compared, for example, with those arising from our enhanced awareness and explication of ecological systems. It is an adventure we are surely just beginning; perhaps advancing our oracular capacities may help point us toward the future. Throughout the course of human existence, our species has learned to use oracular techniques as part of building and sustaining civilisation. Now in our contemporary world as our civilizations now faces great challenges to continued existence, especially in relationship with our environment, perhaps we may find ways to recover and adapt these noetic states of mind in new and creative ways.

References

Adamatzky, A., Armstrong, R., Jones, J. and Gunji, Y.P. (2013). On creativity of slime mould. *International Journal of General Systems*. http://dx.doi.org/10.1080/03081079.2013.776206.

Aristotle. n.d. Available at: http://classics.mit.edu/Aristotle/athenian_const.1.1.html.

Bollas, C. (1987). *The shadow of the object: Psychoanalysis of the unthought known*. London, UK: Free Association Books.

Cambray, J. (2002). Synchronicity and emergence. *American Imago*, 59(4), pp. 409–434.

———. (2009). *Synchronicity: Nature & psyche in an interconnected universe*. College Station, TX: Texas A & M University Press.

———. (2011). Moments of complexity and enigmatic action: A Jungian view of the therapeutic field. *Journal of Analytical Psychology*, 56(2), pp. 296–309.

———. (2017). Darkness in the contemporary scientific imagination and its implications. *International Journal of Transpersonal Studies*, 35(2), pp. 75–87.

———. (2019). Enlightenment and individuation: Syncretism, synchronicity and beyond. *Journal of Analytical Psychology*, 64(1), pp. 53–72.

Coleman, W. (2016). *The emergence of the symbolic imagination*. New Orleans: Spring Journal Books.

du Sautoy. n.d. Marcus. Available at: www.youtube.com/watch?v=sDXMRN2IZq4.

Ferro, A. and Basile, R. (2009). *The analytic field: A clinical concept*. London: Karnac.

Frankel, J. (2011). The analytic state of consciousness as a form of play and a foundational transference. *International Journal of Psychoanalysis*, 92(6), pp. 1411–1436.

Harari, Y.N. (2017). *Homo Deus: A brief history of tomorrow*. New York: Harper Collins.

Hogenson, G.B. (2005). The self, the symbolic and synchronicity: Virtual realities in the emergence of the psyche. *Journal of Analytical Psychology*, 50(3), pp. 271–284.

———. (2009). Synchronicity and moments of meeting. *Journal of Analytical Psychology*, 54(2), pp. 183–197.

———. (2019). The controversy around the concept of archetypes. *Journal of Analytical Psychology*, 64(5), pp. 682–700.

James, W. (1902/2011). *The varieties of religious experience*. Digreads.com. Available at: www.digireads.com.

Johnston, S.I. (2008). *Ancient Greek divination*. West Sussex, UK: Wiley-Blackwell.

Jung, C.G. (1953–77). *Except where indicated, references are by volume and paragraph number to the collected works of C. G. Jung*. 20 vol., ed. by H. Read, M. Fordham, and G. Adler, trans. by R.F.C. Hull. London: Routledge and Princeton NJ: Princeton University Press.

———. (2009). *The red book*, ed. by S. Shamdasani. New York and London: W.W. Norton & Company.

Knox, J. (2003). *Archetype, attachment, analysis*. Hove and New York: Brunner-Routledge.

Le Shan, L. (2012). *The medium, the mystic, and the physicist*. New York: Helios Press.

Leach, M., ed. (1972). *Funk & Wagnalls standard dictionary of folklore, mythology, and legend*. New York: Funk & Wagnalls.

Maul, S.M. (2018). *The art of divination in the ancient near east: Reading the signs of Heaven and earth*, trans. by B. McNeil and A.J. Edmonds. Waco, TX: Baylor University Press.

Martin-Vallas, F. (2008). The transferential chimera II: Some theoretical considerations. *Journal of Analytical Psychology*, 53(1), pp. 37–59.

Merchant, J. (2019). An emergent/developmental model of archetype. *Journal of Analytical Psychology*, 64(5), pp. 701–719.

Ogden, T. (1997). *Reverie and interpretation*. Northvale, NJ: Jason Aronson.

Pollan, M. (2018). *How to change your mind: What the new science of psychedelics teaches us about consciousness, dying, addiction, depression, and transcendence*. New York: Penguin Press.

Powers, R. (2019). *The overstory: A novel*. New York: W.W. Norton & Company.

Ratter, B.M.W. (2013). Surprise and uncertainty: framing regional geohazards in the theory of complexity. *Humanities* 2(1), 119. https://doi.org/10.3390/h2010001.

Simard, S., Beiler, K., Bingham, M., Deslippe, J., Philip, L. and Teste, F. (2012). Mycorrhizal networks: Mechanisms, ecology and modelling. *Fungal Biology Reviews*, 26(1), pp. 39–60.

St. Clair, N. n.d. Available at: www.youtube.com/watch?v=PMerSm2ToFY.

Stern, D. (2004). *The present moment in psychotherapy and everyday life (Norton Series on Interpersonal Neurobiology)*. New York: W.W. Norton & Company.

Stern, D., Sander, L.W., Nahumn, J.P., Harrison, A.M., Lyons-Ruth, K., Morgan, A.C., Bruschweiler-Stern, N. and Tronick, E.Z. (1998). Non-interpretive mechanisms in psychoanalytic therapy: The something more than interpretation. *International Journal of Psychoanalysis,* 79, pp. 908–921.

Strachey, J. (1955). *The standard edition of the complete psychological works of Sigmund Freud, volume X (1909): Two case histories ('Little Hans' and the 'Rat Man')*, Vol. 10. London: The Hogarth Press.

Struck, P. (2016). *Divination and human nature: A cognitive history of intuition in classical antiquity*. Princeton, NJ: Princeton University Press.

Taleb, N.N. (2010). *The black swan*. New York: Random House.

Tero, A., Takagi, S., Saigusa, T., Ito, K., Bebber, D.P., Fricker, M.D., Yumiki, K., Kobayashi, R. and Nakagaki, T. (2010). Rules for biologically inspired adaptive network design. *Science*, 327(5964), p. 439.

Tronick, E. (2005). Why is connection with others so critical? The formation of dyadic states of consciousness and the expansion of individuals' states of consciousness: Coherence governed selection and the co-creation of meaning out of messy meaning making. In: J. Nadel and D. Muir, eds., *Emotional development: Recent research advances*. New York: Oxford University Press, pp. 293–315.

Wohlleben, P. (2016). *The hidden life of trees: What they feel, how they communicate – Discoveries from a secret world*. Vancouver, Canada: Greystone Books.

Woodbury, Z. (2019). Climate trauma: Towards a new taxonomy of trauma. *Ecopsychology*, 11(1), pp. 1–8.

Chapter 13

Persephone's suicide

Craig San Roque

> The first angel sounded . . . and all the green grass was burnt up.
>
> (Revelation 8:7)

Archipelago

He charted a series of dreams. Dreams laid as islands are laid, linked beneath the surface; islands hardly noticed if the mist is up, or the sea has risen. Like old maps those visions are inscribed in his journal. Drawings of strange things encountered, voices heard in a language not familiar, a record of a voyage through archipelago quaking with events in the widening world, as he, in 1913, felt the quaking.

From October 1, 1913 through to July 1914, the prison camp doctor continues his journal, the one which will become known as the *Red Book*. He describes premonitions. He sees floods. He foresees 'the death of thousands'. He dreams of a 'sea of blood covering the northern lands', 'a dead hero', 'a giant foot stepping on a city'. He dreams of 'murder and bloody cruelty', 'a procession of dead multitudes'. He records a voice saying to him 'this will become real'. He describes his soul coming up from the depths asking, 'Will you accept war and destruction?' She shows him further destruction – military weapons, human remains, sunken ships, destroyed states. The sacrificed fall left and right. Then, in June 1914, he has three dreams, the main themes being: himself in a foreign land. Having to return quickly by ship. The descent upon the land of an icy cold (Jung, 2009, pp. 200–21, Shamdasani's Introduction).

The red covered journal of Carl Gustav Jung (2009) is famous enough now, though not as famous as Captain Cook's journal of the Pacific voyages, 1768–1779, nor the 1839 *Voyage of the Beagle* by Charles Darwin. The lands boarded by Cook, the lands observed by Darwin have fallen into dire straits. Of this I can speak because I live upon lands occupied in 1788 by Great Britain. Lands cut open to the world. Each day, where I live in Australia, I see the consequences of that vivisection. Is 'vivisection' too strong a word? Perhaps I should add 'anaesthesia' because the 230-year operation which allows me to dwell here has been conducted while most of us were under a state of anaesthesia. No attention to the effects and consequences of the operation.

The dream turmoil of Dr. Jung was going on a year after he published *The Psychology of the Unconscious*. He was fine-tuned to observe the intensities of unconscious mental processes. Also, Jung and Freud were having a bad time handling each other. Jung confessed that he might be in a psychotic breakdown, the catastrophic dreams being symptoms of personal disintegration; yet, we know in retrospect that the integrity of Europe was heading into collective breakdown. You might say the psychotherapist was himself in the thick of things. I mean he was, like us, infected by the psychosis of the time. Jung had been training himself to observe subliminal personal reactions when in the presence of patients. I see him, in the journal, extending those psychotherapeutic sensitivities into the context of gathering storms of war; catching the infection of the time, attempting to describe the horror of the ignored fact. Jung was a weatherman; though, like Dante, lost in a dark wood, until figures appeared, as Dante's guide appeared, at the bidding of Beatrice, to guide Dante through the Gates of Hell and then on.

Jung's premonitions were not wrong. A full-scale collective psychosis did come flooding with icy cold upon those gathered in Europe; some, indeed, coming in ships from far countries. A 'procession of dead multitudes' did walk into the conflagration; as no doubt did the ancestors of many of us who gaze again upon the brinkmanship of our own times – the melting ice, the silting river, the woman who dreams of seas swallowing her island home – the woman whose life is scattered among ruins, the human flow, the destruction of cities and archipelago.

Whom does the reef serve?

Down the northeast coast of Australia, a long string of coral and vigorous marine life mediates between ocean and land. Strings of pulsating islands run north along Australia's Cape York and through the Torres Strait, Micronesia, Melanesia, Indonesia and the Pacific. This is the region of the sentient archipelago – delicate strings of sand and coral worrying, in their geo-physical kind of way; anxious about swamping, rising tides of plastic and the strange winds of human indifference. Land/sea care is as necessary as care among human beings – two hands of therapy. I think of this action as the 'Persephone Impulse' – simultaneous care of sentient souls and care of ecosystems, interwoven.

The Australian Great Barrier Reef, in ecological fact, is not really a 'barrier' but a permeable responsive, uterine, fertile crescent nurturing an extraordinary spectrum of sea life. The reef is an 'increase site' – a place emanating continuous, seasonal organisms, a seed bed of marine creation. In an indigenous sense a site has to be 'sung' in ceremonies to keep its vitality healthy and productive. 'Singing' means giving tender attention. Presence of mind. To protect the reef from coral bleaching, predatory invasions and the outflow of coastal coal mining the government has to be convinced that the Great Barrier Reef is a tourist business 'increase site', financially equivalent to income from mining. As though money is the sole judge of the fate of the sea, the sole singer. What is this strange slip in the brains of men which keeps one eye open and shuts the other? Whom does the reef serve?

The Australian Institute of Marine Science does sustain projects to restore the Great Reef ecosystem (see www.gbrrestoration.org.webloc). Such people do 'sing' to the reef. Reef therapists are at work, and further north in the Indonesian region, on the north tip of Bali, the people of a small coastal community of Permuteran courageously blocked pirate fishermen from dynamiting their reef, then, with simple technology, set electromagnetic wire nets into the ruined wounds to stimulate a faster regrowth of coral (see www.biorock-indonesia. co#918E17). I have seen this in action. It is simple and, so far, effective. The logic? Coral comes back, the fish return, scuba diving tourists return and money comes back to the village. The villagers think ecologically and economically. Both sides of the brain. And some people can think long term, for generations to come; as does Naomi Klein in her activist video on the Great Reef, made with her young son, for the reef serves the children to come; children of the human, children of the fish. As Klein eloquently declares – 'climate change is intergenerational theft' (see www.theguardian.com/environment/video/2016/nov/07/ naomi-klein-at-the-great-barrier-reef-under-the-surface).

Talking Dugong

Switzerland, internment camp, 1913, the psychiatrist falls into conversation with imagined characters – his soul, in female voice, the prophet Isaiah and a white-bearded elder of gravity whom he names Philemon. There is also the famous young dancing woman, Salome, who helped John the Baptist lose his head. Jung draws intricate patterns, writes in old-fashioned script, imagines while under the influence of the Bible and Northern European imagination – as you might expect from a person steeped in the cultural heritage of his people, archaic visions playing upon the mind. Giving form to his anxieties. There is nothing particularly mad about vision figures coming to speak. Mythologies seethe with vision figures appearing from the matrix of creation and destruction. Myth and fantasy seem to direct human actions as much as anything does. I wonder: what unconscious fantasy drives my response to climate change anxiety? (This reflective question is the underground purpose of this chapter.)

Ancestral beings are place-specific (mostly), communicating in local tongues.

In the Torres Strait islands, Papua New Guinea and central Australia, ancestors talk in the iconography of their country – the surrounding sea, the rippling desert. Cultures and languages mix up; that is so – in my region, there is much tumult in the psycho-environmental world. I imagine that there might be tumult among the ancestral creation beings; perhaps no being knowing quite what to say to whom, now.

If some ancestor came to speak about the present problem of strange things happening in the desert country around Alice Springs (unprecedented fire, water pollution, feral animals, voracious weeds, relentless fracking), the ancestral creation being might be a mixed up eagle-winged/lightning flash/burning tree Dreaming being, worried about white people doing blind-eye things to precious country.

Or if He/She worries about the sea swallowing islands under the tropic of Capricorn, the local Dreaming being would most likely be a turtle, or maybe a dugong, mixed up with plastic bags, talking Dugong.

Dugong are a marine mammal living in warm coastal waters of the Indian and Pacific oceans. Dugong feeds on seagrass. Dugong eat a lot of seagrass, but the seagrass doesn't seem to mind. Dugong and seagrass have come to an arrangement. Australia hosts a large population of dugong, especially along the northern coasts. The dugong features in coastal indigenous lore. There are dugong stories and dugong songs sung in dugong ceremonies. Dugong would not feel at home in Zurich, even in the lake.

Dr. Jung did not speak with Dugong in his *Red Book* reverie; he opened an ear to his European ancestries, his local stories. He heard trouble. He saw a flood of blood sweeping across the body of disintegrating Europe. He did not deny what he saw, what he feared. He set about procedures of diagnosis.

Gulf of Carpentaria

> And they that are with him are called and chosen and faithful.
> (Revelation 17:14)

I sometimes read the Bible. I like Genesis, Isaiah, the physician Luke and John of Patmos for his Revelations. I like the way God speaks now and then or goes up in a pillar of smoke. Angels knock at the door. I do not like the story of a chosen people. I do not like the story of a special human dying to make me welcome in heaven. This, I think, is an ontological trap feeding a perverse idea that a body has to be torn apart to secure salvation and peace on earth. I wouldn't worry about primitive sacrificial fantasy, except I notice the current Secretary of State of the United States (Mike Pompeo) seems to believe in bullet holes in the sky – looking forward to the 'Rapture' when the world ends and he, as a member of the chosen congregation, will ascend through those bullet holes to the bosom of Abraham. This, I think, is how undercover fantasies direct the politics of earthly care.

Today I read Alexis Wright. Alexis writes a book of revelation for my country. She writes in the language of tidal fish, serpent, raincloud; the vast sorceric heat of Australian indigenous body-thinking. She is no John of Patmos in seclusion on a Greek island. Alexis Wright is absorbed by movements within her country, movements of spirit and the facts of our present condition. Her apocalyptic visions emanate from the deep of local experience. Her unique voice is recognised. Her books gather distinguished literary awards. Her writing mind is resolutely of Australian multi-cultural descent. Her mother's ancestral Aboriginal country borders the Gulf of Carpentaria where her novel *Carpentaria* (Wright, 2006) is set.

On her mother's side, Alexis also traces Chinese migratory connections. Her father was a white man who worked the cattle in the north of Australia. Her sensibilities are, therefore, woven from a rich genetic basket; hybrid, resilient, wide open and fractal. Among such people, you find those most concerned with regeneration.

Carpentaria rolls in the language of indigenous suffused Australia – multi-racial characters, multi-cultural cosmologies, deranged white and black men/women, ambidextrous kids subsisting in broken houses and uncertain times. Nightmare images slide in out of perspective, as in Jung's 1913 psychosis visions. If Jung's psychic confusions reflected Europe in his time, Alexis Wright's *Carpentaria* reflects hallucinogenic confusions in the Australian collective brain; Aboriginal dream thought mixing with bitumen, sand shoes and frothing British shampoo. Nothing is really clear or clean. Broken images submerge, toss and tear through floods of humid magical thinking, steeped in the cultural imagination of that intricate, embodied presence known, in popular English, as the 'Dreaming'. The Biblical book of *Revelation*, Dante's *Divine Comedy*, Jung's *Red Book* and Wright's *Carpentaria* are all of a piece as works of visionary imagination responding to catastrophe, seeking regeneration, mercy, hope.

I wonder could Captain Cook have envisioned what would follow from his discoveries? Could Darwin have imagined what would evolve in the Pacific after he wrote his *Origin of the Species*? Today, perhaps, Darwin is writing *Death of the Species*. OK. I can accept the facts of the shifts in our earthly environment. I can accept there must be a spectrum of psychological reactions swirling in the minds of people in response to environmental disorder: images of panic, depression, paralysis, last minute lust, predation, denial, survivor humour? Who knows? The 'climate disorder psychological reaction spectrum' would make a *Diagnostic Manual* of its own.

And I wonder about some strange dissociation, the forgetting of the evolutionary chain of being through which we descend – a chain of living, evolving consciousness in which we so obviously participate. Perhaps this is the fundamental dissociation; the denial of origin, the failure to care for origin. Each day I eat the plant and animal bodies of those origins, knowing I come from all this, yet I do so little to nurture, pragmatically, the continuity of waters, lands and species which sustain us. Each day we fell ancient trees. As each habitat falls, a neuronal link in the great chain of being falls.

Blue crab dreaming

My granddaughter and I are walking on a beach, on the mud flats at low tide. Swarms of little blue soldier crab running before us. They bury myriad little feet, slipping into wet tidal sand. They disappear. Gone in a trice.

> 'Look', she said. 'When they grow up, those blue crabs will be thoughts in somebody's brain'.
> 'How is that?'
> 'Those little blue crabs are making somebody's brain in the future. Millions of thoughts running along together'"

'You mean, the beach is thinking'.

'Yes, then the thoughts bury in the sand. All gone. But don't worry; thoughts come up again when the beach is quiet'.

'What would happen if this beach and all the crabs died?'

'Well', she said, 'There'd be no more brains in the future'.

From the movement of creatures, we learn patterns of creation, the voices and the pattering of crab feet in the sand. Observation builds the repertoire of human thinking. A human body holds the history of the long becoming. In my country, the acts of nature observed are woven into long song lines describing the movements of pragmatic survival, of life/death and beings of imagination. The weaving net of such facts of life are named in Australia as *Tjukurrpa, Altjerre, Wongar* . . . many names for the creation stories interlaced throughout the continent. Actions of regeneration ceremony wrapped in several hundred indigenous languages. Languages. Cradles of civilisation.

I imagine you with me on the mud flats with my granddaughter, wading through the scattering thinking blue crab. Becoming attuned. I will tell you a Dreaming story from the Gulf of Carpentaria. Then we can return to Alexis.

Two women, baskets and two men

Before people crossed the seas into the Australian continent, a long time ago, the way was prepared by four great towering beings – the Djanggawul sisters, their brother and his friend. What I tell is only the tip of this tale. Ronald Berndt (1953) made the English translation from the original Yolngu verses in this marvellous song cycle from North East Arnhem Land.

The story is not to be chattered over coffee in the city. OK? This is a story about serious matters. It tells how creation beings crossed into Australia in the beginning, bringing plants and children to life. I had it from my father-in-law, who had it from Berndt, so I guess I feel a family connection.

It is said the Djanggawul beings crossed the sea into northern Australia, coming south from Bralgu, the spirit Isle of the Dead – so it is said. They came in over the waves by canoe, from the north, following the path of the Morning Star. The Djanggawul sisters carry woven dillybag baskets full of fertile power. They carry the basket (*nganmar*), full of children. Yes, and the basket full of male fertile power. So it is said.

Look, goes the song . . . the Djanggawul brother has left the canoe, he is wading to the beach. He rises from the water at Blue Mud Bay, his face foam stained, his body patterned with salt water marks. The sisters are singing – Djanggawul look back and see the rays of light leading back to our island of Bralgu.

Shine that falls on the paddles as its dipped into and drawn from the sea. Shine that spreads from the star rays from Bralgu. . . . The Morning Star skimming the sea's surface. . . . Foam and bubbles rise to the seas' surface. A large wave carries us on the crest. . . . The roar of the sea . . . its salty smell.

The Djanggawul come ashore. They carry sticks. Where the brother thrusts his stick into the ground, water wells up. Where the sisters thrust their sticks into the ground, water flows. Long straight slender trees grow. The women, pregnant always by the Djanggawul brother, gather children from the woven womb baskets they carry. Those children will be the people to come. Those people who live in that country came from the Djanggawul. They came from Bralgu, following the Morning Star. Djanggawul left animals and plants for the people. Everything comes from the Dreaming – do you hear?

Such things are told, accounting for the management of creation and relations between living things. A ceremony/story like Djanggawul is history and the basis for cultural human law. Dreaming trains, the mind of the people. Most white people don't realise this. Aboriginal people, where I live, say: when people forget Dreaming, the country gets lonely. Country becomes depressed, as well it may. Nobody listens for blue crabs thinking.

Cyclone

In *Carpentaria*, Chapter 10, a great cyclone swings down from the tropical north, hitting the Gulf roundabout where the Djanggawul travelled in, then rolling south like some giant cloud enshrouded being. The seasonal weather change, the monsoonal 'wet season', overshadowing.

> if you were to see miracles happen look to the heavens in November . . . look for the giant in a cloak. Brace yourself when he comes rolling through the dust storm, spreading himself red, straight across those ancient dry plains heading for town. . . . That's right. The giant *sugarbagman* of the skies walked from horizon to horizon carrying storms and hazes of madness, and sweat. . . . People went mad from it; Uptown people called the *sugarbagman* spirit "seasonal rains" or their "silly season" and among them were fatalities. Statistics rocketed in mortality for both black and white.
>
> (Wright, 2006, p. 308)

Jung in Switzerland, 1913, foresaw mortality statistics rocket and he, too, saw strangely cloaked beings walking horizon to horizon, raining mud and blood, picturing the whole mess like some giant Grendel Gotterdamerung cannibal mother or Wotan in uniform. That's a Swiss German thing to do. In north Australia, 'The Wet' comes every year and people get over it; but this time, in *Carpentaria*, 'The Wet' gives way to a monstrous cyclone which rips apart the Gulf Country, swallowing towns and people. The serpentine cyclone arrives smashing, flooding the coastal town and camps of Desperance, where Alexis Wright's 'hero' Will Phantom and his Aboriginal family live in rusted corrugated iron shacks. All are caught in the floodtide.

> what a catastrophic requiem took place in those floodwaters racing out to sea . . . the waters poured dead fish. Sodden spinifex grasses. Sticks. Green

wood. Branches. Plastic. Plastic Malanda bottles. Green bags tied up with rubbish. He drank the stinking air manufactured by the porridge of decaying fish, and gladly, the nauseating stench touched him. . . . In the mayhem of buoyant bodies, bloated animals, floating by touched him ever so lightly . . . other things touched him too, and the madness went on and on . . . he was astonished then weakened by the feeling of helplessness . . . he felt like he was an intruder to be clinging to a foetus inside the birth canal, listening to it, witnessing the journey of creation in the throes of a watery birth.

(ibid., pp. 492–4)

Will Phantom, tumbled in cyclonic debris, dragged through muck and water-logged detritus, thinking of the tumult as a great slippery serpent; as well he might, because in Will's indigenous mind, the sea, rivers and lands are animated, suffused with spirit creatures whose power is always dangerous: like the giant in the cloak of storm cloud; like Djanggawul, like those northern archaic beings who stalk the realms of Europe.

For months, Will lives marooned on a floating island of plastic bottles, fishing nets, broken trees, becalmed in the Gulf of Carpentaria, a lone survivor. His beloved wife, poignantly named Hope, and their only child are dead or lost. Lost Hope. Hope has been captured by militant security guards from the mining company and dropped from a helicopter into the sea – an act of reprisal against her dissident husband, Will Phantom. In Alexis' story, industrial mine directors are set to dig further into protected Aboriginal land, offering financial compensation to custodians of those Dreaming sites that could be disturbed by the extractions. Disturbance of eco-psychic/significant Dreaming/sacred sites is a constant, controversial story in mineral-rich Australia. Will and Hope, unlike many of their destabilised kin, resolutely refuse to accept the money, initiating a guerrilla resistance against the mining company until the cyclone sweeps in, changing everything.

'Persephone's Mother'

Such things have happened before in all the regions of the world, floods sweeping in changing everything; a volcano, an ice age, fertile lands turning to sand. There are accounts of such matters, stories lodged out back of the human brain. Cultural memories – some tell of hope, resilience, survival. Some do not. 'Persephone's Mother' is such a story from the Mediterranean region. This is how I tell it.

A mother and daughter travel across the beginning of time. The young woman walks in the company of a dog (Persephone's dog). The dog dies; by mistake, eating poisonous fungus. The spirit of the dog goes down through a crack in the rock. The young woman follows. She meets a being whose face no living creature sees until the moment of death. The mother looks everywhere, calling, 'Have you seen my daughter?' She is angry. Everywhere she walks, she blasts the earth with cold. Everything dies. This is mother's grief, the people say; we can do nothing about this. The young woman is with the serpentine being. She travels with him.

He tells her, 'your mother thinks I've stolen you. Pity. I'm showing you the roots of things. Death and life twine together. Remember this'.

They travelled the world. They travelled the sites of the world. 'Look', he said, 'here are dangerous places. Savage things happen. Here creatures fall to death. Teach the people to take great care. This is the way of things. Remember this'.

That no-name serpentine being, he moves underground, wrapped in earthquakes. He deals with everything found underground. He deals with the dead. People fear him. People will not look him in the eye. The young woman comes to love him. She looks at him with care. She comes to understand the nature of things, underground. She comes to understand that no-name serpentine thing. She names him – Aidos/Hades – 'keeper of the dark, keeper of the dead'. Now the souls of all the dying things fall to her hands. She gathers them like seeds into her basket. She takes care of them. She handles them like clay. She makes them human. She is named – 'she who is bright yet wedded to the dark': Persephone.

Persephone returns. Her mother has thrown the living world into endless night. Nothing grows, no birds sing. The people say – this is mother's grief, we can do nothing about it. Our only hope is when her daughter returns. The daughter returning from Aidos cannot tell the full truth of what she has seen, what she has become. She makes up a story. The mother cheers up. The dead weight of fear is lifting. People are fed. So goes the old story. Seeds, plants, fruit come back to life. People sing, telling children how such things happened. Summer is in full light. Autumn approaches; cold winds sweep in. Things change. Persephone walks through that crack in the rock. 'Don't worry', the young woman says. 'I'll be back again in the spring'. Perhaps. Perhaps.

Radio Eleusis

You may recognise the characters – Aidos/Hades, associated with death yet also, at the partition of the primal powers, he became custodian of riches below ground; soil, root, mineral, oil, titanium, uranium, coal. The mother, Demeter, became custodial generator of flowing nature. She, like the Djanggawul sisters, crossed a sea, it is said, bringing from Crete the fruit, seeds and cultural lore of cultivation. Her offspring Kore/Persephone is she who descends and returns – chrysalis, beehive, barley, wheat, sister of the vine, icon of organic cycles embedded in the work of the land. Regenerating.

You may recognise the myth of Demeter's grief as a post hunter-gatherer account of confusion in the natural order, ice age onslaught, floods, shifting environment, adaptation, migration, digging, planting, reaping, sowing; versions told and retold along the trade routes of the Fertile Crescent – through the Mediterranean, Egypt, Syria, Iraq, Anatolia, the Caucasus regions; crystallised at the sites of Eleusis, Greece and Enna, Sicily. Stories from the northern *homo sapien* 'coming into being' – primal organic women, blood-spilt men, doing what they must, carrying secret/sacred things – blood lines, seeds – as did the Djanggawul, the ones who came ashore.

You remember back in 1913 that doctor camped on the brink of war, and that soul thing rose up in the night talking to him in his own language saying, 'Will you accept this destruction?' Maybe there comes a moment when she rises up to each one of us, talking our own language, 'Will you accept this destruction?' When it happened for me, when she did rise up in the night, she wasn't talking Dugong or Blue Crab – she was talking 'Persephone'. The situation, the lucid Dreaming story which wrapped around me, was the funeral wake of Demeter's daughter. *Persephone's Wake*. Her suicide in fact. Persephone leaving. It came as a shock that the creation being who, for so long, embodied the return of fertility might depart. She would not come back. Persephone brought to her knees, giving up her job, unable to accept the destruction which we, the people, have wrought. Persephone leaving for the last time. This story is broadcast from Radio Eleusis.

Persephone's Wake

Persephone's Wake (in development) is part three of a trilogy of community theatre works. *Persephone's Dog* was performed on a broken cliff face outside Alice Springs in 2015. It offered an alternative ecologically attuned version of her descent and return from the Underworld. *Persephone's Heart* (performed in 2016 in Alice Springs, then in Santorini in 2017) tells how and why the 'regenerative divinity' acquired a functioning human body with the help of her dog, the Seven Sisters, and specific geographic sites and kin in ancient Australia (San Roque, 2015, 2019).

Persephone's Wake tells how and why Persephone gave up her body and function. You will hear references to the Sumerian descent/return myth of Inanna (3,000–4,000 BCE) set in southern Iraq and the city of Uruk (Wolkstein and Kramer, 1983). Inanna's execution and revival in the Underworld (The Great Below) featured in *Persephone's Heart*. Those events are recalled in my dream-like *Wake*, Persephone appearing in much the same way as mysterious figures appeared to Jung when he was in the thick of things in 1914 – as are we now, in our time.

The site of the wake is a barren claypan surrounded by dying trees. People camping. Fires smoke. A body, Persephone, wrapped in a woven net of grass and branches is suspended from a tree. This is a burial custom, the body left high to decompose in the company of birds. The family of Persephone is gathered, mingling with birds, dogs, musicians. Persephone's daughter Kore, her mother, Demeter, all silent, sitting with Hekate/Crow and family, faces smeared with white ochre. The wake has been going on through sunset and into the night. We have heard stories of Persephone's life, her travels, her wedding, her work. Now all is quiet. Hades, who is not seen, is present within a black rock within a fire tended by Persephone's son Kouros and his cousins. Kouros arrived late, having come from a city in Iraq – the site of the ancient city of Ninevah (Mosul), where Kouros works in a war casualty hospital.

Night. A lament swells through the camp. The spirit of Persephone appears. She walks through the fires to her mother. She kisses her. Her mother does not move, saying only, 'Why don't you come home?' Her mother turns to stone. Persephone, touching her mother gently, moves to her son and daughter, caressing tenderly their faces. She speaks.

Persephone

Kouros, I have seen your city of Nineveh. I have seen every city since the beginning. I have held souls in every broken city. They fall as rain. Nothing grows from such killing. Who are these men who govern the fall of the world? They betray the women who gave them life. How is this that women make such men? Fire every morning, fire every night. I hear them say – burn the land and you will have food. No, I say – the best land is land held in my mother's hand. Bless the bees who turn the earth.

Kore, you remember. In the beginning, my mother and I walked the earth. We gathered seeds for the people yet to come. I named the plants. We cared for them. Now I see what such people have become. Let them fall. This is how I speak. Remember this.

Kouros, speak to your sister. Tell her I love her. Tell her to remember the love I found with you all. Tell her to remember the beehive. This is how I speak. So much the love. So much the earth. Always the light and shade.

Do you hear?

Kore

Hold your hands over the floods; hold me, hold me. Do you hear?
I love you, do you hear? I remember the beehive.
I remember the names of all the plants.
I remember the basket held by you. How can you do this?
How can you leave?
How can you not come back? Do you hear? Do you hear?

Persephone

These hands torn by hungry dogs; how can I lift these hands again Kore?

My mother told me the people would come. 'Teach them to make bread', she said, 'Teach them the plants'. I waited for the people to come. In the beginning, I saw them walking, hungry, looking for food. I helped them. To those people I say, 'I see what you have become. You sing no lullaby for the growing thing. Do you hear? I fed you. I caught you falling. You remember? You thanked me. We were bound together in a promise of love. Now I see your men wake in the morning. They go out and burn. I see women blind'; 'Burn the earth' – they cry – Dress us in silver and silk and burn the earth'. Do you see? Do you see what you have become?

I curse you now and at the hour of your birth. Do you hear? May the hair of such women fall to the ground. May the teeth of such women fall to the pit. The hands of such women have turned to hooks. Do you see? Do you see? Kore, I saw what happened to my sister Inanna, hung in Sumer upon that hook. I told you what happened to my sister Inanna beneath Uruk. You have seen that hook.

Do you hear? Do you see now? This is my body hanging on that hook.

Kore, look at me now. What have I become – what do you see?

Kore

I love you. Do you hear me? I see your lungs choking with cities.
I see your liver choking with poison.
I see every cell in your body meeting death.
I see the nerves of your body broken. I see your beauty stolen.

(Choking, shaking, sinking to the ground.)

Persephone

And this, I say this – do you hear me?
On this day. On this night; I, Persephone, Queen of earthly things,
I, Persephone, Queen of Death and growing things;
I, Persephone, before you all. My shoulders fall. My arms fall.
My hands turn to salt. My feet turn to salt.
My most sacred places fall to dust.

Do you hear me? I love you.

So full of bees. So full of flowers. All the little birds are falling.
Nineveh and Jerusalem never saw such a fall as this.

(Persephone fades)

Hekate

(Coming to Kore, with a blindfold, she ties the blindfold around Kore's eyes.)

Kore, can you hear me? Remember, your mother was always a cicada, underground for a long time. She came out. She sang. She was fertile. She gave birth. She left her beautiful shell. Where has she gone, Kore? What has become of her? Until you know, do not despair. Do you hear?

Follow with the eye I gave you. The secret eye I gave you when you were born. Remember? Follow her. Remember the seven gates of Inanna. Remember the country of Sumer. Remember your mother went down to meet her sister Inanna so long ago. Remember those seven gates to the Great Below. Go down through the seven gates. You will find the track of your mother. Will you do that?

Kore

If not me, then who? If not now, then when?
I love her. Do you hear me?

Hekate

(Fixing the blindfold)

What do you see, Kore? What do you see?

Kore

I see my mother in a cave – I smell fear and dark. No, not in a cave. It is a hive. I smell the honey. I see my mother in the beehive. I hear the humming. She is inside the beehive. The caves so full of honey. Humming. Do you hear?

Persephone's voice

Kore, do you hear me? Do you remember?
Each time I came back to you, I passed through those seven gates.
Each time I went down, I lost all at those seven gates.
Each time I come back they ask for my identity.
Who am I? I say, 'I am the life that keeps your women alive'.
The creature at the gate, in his uniform and hat, he laughs at me –
'It's my dick that keeps the women alive'. I look at him and spit.

I say, 'You know nothing – your brain is smaller than your balls,
and your balls are smaller than a frog's arse'.
They beat me again. They do that each time I come back.
Do you think I walk back through a corridor and everything is light?
At every step a mother begs me to bring a child back to life.
How can I do that? How can I look that mother in the face and say 'I cannot
  do that'? I love you. Do you hear me? But I cannot bring a child to life.

Kore

I can hear you. If not me, who? I am at the gates . . . going down. I am following. I see your steps. Layers of dust, how many thousand years of dust, and
I see your footprints, going down. Over and over.

Hekate

What do you hear?

Kore

Do people think the Morning Star rises without pain? Do people think the
evening star sets without grief? Do people know the pain it takes to keep this
earth alive?

I am following her footsteps. I am going down. My belly and her belly, my
heart and her heart. I am going down. Where is my brother? He should be
with me. Going down.

*(Hekate calling Kouros wraps the black blindfold around his eyes. He 'steps' into
the dark)*

Hekate

Kouros, what do you see?

Kouros

I see my grandmother in a rage. She is tearing her hair; she is ripping her
breast. Everything is undone.

Hekate

You see the present time Kouros. And now, what do you hear?

Kouros

> I hear wild dogs howling.

Hekate

> And now, what do you see?

Kouros

> I see my father. No. He is gone. He changes. He shifts. I can hear him. He is
>   whistling me. There is black smoke all around him, air burning.
> I keep walking. Walking on broken ground, walking through ruin.
> I see burned out tanks. I see a gate. I am passing a checkpoint.
> I know that checkpoint. The Ninevah Gate. The Euphrates crossing.
> They stop me, they ask my identity.
> Who am I? I am Hari Kouros, son of Aidos.
> Occupation? I walk through minefields.
> No not a soldier. I am not armed.
> I pick up the pieces.
> Address? The hospital in Nineveh.
> Walk on. Your father is waiting.

*(Kore, blind, her hands outstretched. She is walking through a crack in a wall, a
tunnel under the city.)*

Kouros

> Kore, where are you? Do you hear me?

Kore

> I am at the final gate. I hear the bees. The guards take my arms. My
>   identity? I am laughing at them. I say, 'I am an oasis in a barren land.
>   I am running with water'. They laugh at me.
> I say, 'The oasis is always open. Anything you want from this body, I will
>   give it'. They must think I am crazy.
> Who am I? I am Kore, daughter of Persephone.
> Occupation? Oasis.
> Address? Oasis.
> Walk on Oasis. Your mother is waiting.
> They spit on my back.
> Now. Now. Everything confused. I cannot see, I cannot think.

I pass through the final gate. I see ruined ground in all directions.
I see smoke. I see my father. Yes. He is carrying the body of my mother.
He is walking across the country carrying her.
I see my father walking. What is this?

Kouros

I see cities burning. I see towers. Broken. I see smoke.
He carries her. They appear. They disappear. Her body is breaking.
Falling, he cannot hold her. What is this?
The Underworld has come to the surface of the earth. There is no Great
    Below. There is no Great Above. There is only this.

Kore

I followed my mother down. I passed through the seven gates.
It was not as I had been taught. I stepped through the last gate.
This must be the Great Below, I said.
I stepped onto the surface of my own country. What is this?
The Underworld has come to the surface of the earth.
I see them. The smoke shifting. He is carrying the body of my mother.
She is bound together with grasses. She is woven together with seeds.
He is carrying her across the barren ground; seeds fall from her hair.

The seed stock of the first people, falling as he walks. He turns his face to me.
I see his eyes; he is crying. Carrying her seeds, falling; calling for rain. Do
you hear me. I love you. Do you hear me?

Kouros

He cannot do this, Kore. It is too late. She will not come back.

Kore

I don't understand him. He is speaking,
He is saying, 'Help me Kore. Help me'.
Her hands, her feet . . . her body falling as he walks.
I do not want to see this.
Where her body falls, the earth will live again. I am stepping forward –
I will help, I say. I will help. He is trying to replant the world.

Kouros

He cannot do this, Kore. It is too late.

Kore

I hear a voice: 'I say to you, Persephone's child.
Become an oasis for all things. Do not sleep. Keep the oasis open.
Anything love needs from your body, give'.
Why do I speak like this? *(Scratching at the blindfold.)*
How much do I give? Where is this?
How have I come to this? I was feeling for the roots of things. I could hear
   humming. I could hear the bees coming. But what is this?
I see the face of my father. He is crying. My mother has gone.
What to do?

(Kore/Kouros remove blindfolds.)

Kouros

You fed us visions, old crow. Do you hear me? What are we to do? If I had
the seed stock, I would hide it. Keep it ready. The chance might come again.
The bees are all gone. They might come back. What chance, what chance do
we have? *(Pouring sand on the fire, walking away.)*

Hekate

He walks away. Here am I, 'an old crow', and the children walk away.

We are on our knees, crawling in mud. Your grandmother's turned to solid
rock. Your mother gave everything she had. Your father wanders mad, covered in ash. Your brother walks away. And you, girl, you?

Kore

My grandmother might come out of her rock. My father might come back.
My brother will do the practical thing. He has always done so. And I? My
mother's net hangs in that tree. Two eagles flying in; they take her heart, her
liver. They fly to the nest. They feed the chicks. Young eagles will carry the
heart of Persephone. Do you hear me? I love you. Wild dogs come, dragging
her bones across the claypan.

Someday, sometime, someone will put her bones together again. Do you
hear? And as for me, I am here with you tonight, telling you the story. I'll
practice being a new Persephone. What else is there? If not me, then who?
If not now, then when? I love you. Do you hear me? Do you hear? Do you
hear? . . .

A New Persephone

There is a line in Freud's *Mourning and Melancholia* where, writing from Vienna in 1915, reflecting on suicide, he comments, 'we cannot conceive how the ego can consent to its own destruction' (Gay, 1995, p. 588). Yet it seems this is exactly what we do – consent to our own destruction. Naomi Klein (2016), thinking about her children and speaking from the Great Barrier Reef, 101 years later, says – 'Climate change is intergenerational theft'. Good line, Naomi. Maybe your children will become the new Persephone.

Meanwhile, my granddaughter and I walk along the tidal mud, encouraging the blue crabs. 'Keep thinking, crabs', we whisper. 'Keep thinking'.

Acknowledgments

Kore's line, 'if not me, then who? If not now, then when?' are words attributed to Greta Thunberg and young activist groups, e.g. Extinction Rebellion, You Tube, September 12, 2019. Those lines may have originally come from Rabbi Hillel the Elder, born in Babylon, 110 BCE (https://en.wikipedia.org/wiki/Hillel_the_Elder).

Thankyou Odysseas Elytas. Do you hear? Thankyou Uti Kulinjaku group, Central Australia.

References

Berndt, R.M. (1953). *Djanggawul: An aboriginal religious cult of north eastern Arnhem Land*. New York: New York Philosophical Library.

Gay, P. (1995). *The Freud reader*. London: Vintage.

Jung, C.G. (2009). *The red book*, ed. by S. Shamdasani. London: W.W. Norton & Company. (especially the Shamdasani Introduction, pp. 200–237.)

Klein, N. (2016). *Naomi Klein at the great barrier reef – What have we left for our children?* Available at: https://www.theguardian.com/environment/video/2016/nov/07/naomi-klein-at-the-great-barrier-reef-under-the-surface [Accessed Aug. 10, 2020].

San Roque, C. (2015). *Ancient Greece, modern psyche – Archetypes evolving*, ed. by T. Singer et al. London and New York: Routledge.

———. (2019). *Ancient Greece, modern psyche – The soul remembers itself*, ed. by T. Singer, et al. London and New York: Routledge.

'Persephone' performance images. Available at: https://aras.org/newsletters/aras-connec tions-image-and-archetype-2016-issue-2.

Wolkstein, D. and Kramer, S.N. (1983). *Inanna*. New York: Harper & Row.

Wright, A. (2006). *Carpentaria*. Artarmon NSW. Australia: Giramondo.

Chapter 14

'The Singing, Ringing Tree'

Andy White

> You must have shadow and light source both. Listen, and lay your head under the tree of awe.
>
> Rumi (in Barks, 2002, p. 87)

'The Singing, Ringing Tree' is a modern rendering of Grimms' fairy tale, 'Hurle-burlebutz,' in which a naïve Prince calls upon the neighboring King to ask for his daughter's hand in marriage. The Princess treats him rudely and throws his gifts on the floor, saying she will only consider him if he finds the fabled Sing-ing, Ringing Tree, whose whereabouts are long forgotten. The Prince journeys to the furthest reaches of the kingdom where he finds a stone bridge to a secret land guarded by an Evil Dwarf, who captures him. The Prince explains himself and the Dwarf perks up. He has just such a Tree and will give it to the Prince if it sings and rings as proof of the Princess' love by sundown or be made his slave. The Prince foolishly agrees, 'or may I be turned into a Bear,' he adds, which was a rather silly thing to say to an Evil Dwarf whose best thing is a magical challenge.

The Prince returns with the Tree and patiently explains all the Princess needs to do is love him for the Tree to ring and sing. His miscalculation as to how these things work turns him into a Bear. This spell can only be broken by the singing of the Tree which the King has now discovered, having been sent out like a lickspit-tle by his tempestuous and demanding daughter who has changed her mind and wants it after all.

So, our heroine is not very nice to begin with – and why should she be? Her father is weak, yet still treats her like chattel and the stupid Prince thinks he can buy her like a cheap whore. And where is her Mother? Maybe the ugly side of the Princess is what you get when the Queen is squeezed out of the story. The loss of the Mother/Queen in Western culture has given rise to inestimable grief in our time, like the traumas of infancy. It spills from the analytic couch of the few, slides from the slumped shoulders of the many, crumples us all before the blink-ing, blinkering screen. Longing then embeds itself in stuff, creates ruts and gathers clutter, mourning becomes a vague feeling of devaluation or of somehow being unwanted. 'We will do anything to make sure life is secure, even if it is static, rotten and dead' (Woodman, 2009).

To transform this means becoming conscious of the fear of being fully alive, which precipitated it. In the meantime, buried grief appears outside us, banished from persons to stuff, as though the myriad things were like weaving threads in a comfort blanket, magically fending off loss for as long as we surround ourselves with it. Radiant must-haves and bucket lists serve a purpose beyond mere diversion or amusement. They make us feel momentarily whole again.

The King returns to his castle, having promised Bear he can have the first thing he sees when he gets home in exchange for the Tree. Of course, it is the Princess. 'I thought it would be the dog,' he explains sheepishly. No-one dares point out the strange quirk of fate which makes a bitch of his daughter and his noble self in possession of the Tree. Bear arrives to claim his prize, defeating the King's entire guard, abducting the Princess and making good a magical escape.

In the Grimms' original, the story is dominated by this betrayal of the Princess. In the earlier rendering, the King agrees to give his daughter to the Dwarf in exchange for a way out of a forest in which he has become lost. He uses her as a bargaining chip and sacrifices her future in return for his own walled off peace of mind, a dark tradeoff being increasingly protested by groups like Extinction Rebellion who see their futures being likewise compromised by the needs of the powerful few for an easy way out of life's difficulties.

When the Mother/Queen goes missing, the masculine principle becomes split into an ideal yet inflated patriarch on the one hand and an Evil Dwarf on the other, like Jekyll and Hyde or two very different halves of a holy book. So, the Princess cannot help but be at odds with the Prince who, as Bear, represents the ground of her psychological being with whom she is now thrown together by the King's Freudian slip, an 'accident' which reveals his unconscious wish to possess the tree at the Princess' expense.

When they arrive back in the Dwarf's secret kingdom, the narcissistic Princess demands her feather bed, her golden cup and silver plate. But they are all left behind, treasures of entitlement rudely supplanted by mere luscious berries to eat, lousy spring water to drink and horrible soft moss to sleep on. Being thrown together with Bear is a form of shamanic initiation for the Princess. Her betrayal by the King deprives her of the normal reference points of reality. She is thrown back on her own primal resources with which she is yet to broker a relationship. 'When a living organism is cut off from its roots, it loses the connections with the foundations of its existence and must necessarily perish. When that happens recollections of the origins is a matter of life and death' (Jung, CW 8, para. 180).

The Princess' recollection of her deep psychic roots begins with an experience of abduction. In many shamanic traditions the initiate, 'vanishes from the village for months, having been abducted by superhuman shamanic masters. In Nepal, for example, young men are occasionally spirited away by the Yeti, and forced to toil miserably for years' (Mesocosm, 2012). In both Tungus and Udeghe cultures from North East Asia there are stories of women who must marry Bears and go live with them. In Western culture we may not have such helpful stories but can still connect to what it might mean to be seized by an inspiration, grabbed by an idea, or carried away with enthusiasm, the origin of which comes from the Greek 'Enthous,' meaning to be possessed by a god.

In the Viking times which spawned our story, suitably anxious kings had as their immediate magical protectors, Bear-men, Berserkers. The word, 'berserker,' comes from the old Norse, 'ber-sekr' meaning 'bear shirt', which is how these warriors would go into battle, without mail or armor. The purpose was to 'hamask' to change into the Bear itself and tear into the enemy's ranks like beasts. Fortunately, Bear embodies way more than aggressive abduction. He is also an ally whose capacity for relatedness will play a large part in the Princess' difficult journey. She sees him speaking gently with other creatures and demands Bear give her his secret of making animals like him. He says the problem is she appears arrogant, heartless and obstinate to them. She sarcastically wishes to appear as others see her but doesn't realize the Evil Dwarf is listening in and magically makes it so.

The Evil Dwarf is the counterpoint of consciousness seduced by its own self-sufficiency, a trickster figure grown powerful in the absence of the Queen, 'whose chief and most alarming characteristic is his unconsciousness' (Jung, CW 9i., para. 472). Such a boundary-crosser is as comfortable with entanglement as he is with honest conversation. Like Saturn devouring his children, he can gobble up and possess just as readily as catalyze our unfolding story. This ambivalent figure, full of chaotic energy, tends to enter life unannounced and stage left just at the point in which consciousness has become lopsided or narrow in its view, upsetting apple carts and engineering events without too much concern for outcome.

Our Princess is now ugly and disheveled. She flies into a rage but there's no denying her hideous reflection in a nearby pool. This unwitting encounter with her personal shadow immediately expands consciousness. Suddenly, she realizes she can only gain the love of creatures by loving them in the first place. Bear has already learned this by his earlier failed efforts to compel the Princess's affections. She and Bear have something in common. The Princess softens. She calls Bear 'Dear Bear' and says, 'good morning.' They begin to co-operate and build a shelter together, much to the Dwarf's annoyance.

Bear may be dangerous, but he is also a principle of psychological organization once consciousness has corrected its attitude. When such 'a loyalty, or feeling, constellates, it calls forth the secret order which is at the heart of the chaos of the unconscious' (von Franz, 2001, p. 94). This secret order is the author of transformation, much needed when collective attitudes have yet to cross a developmental threshold which might transcend the partisan interests of wealth and power.

In the earliest shamanic Bear cults throughout all Northern cultures, in evidence as long ago as 80,000 BC, the Bear is uniformly recognized as a representative of the gods. In a time when humans and bears shared the same caves, they also shared identity. Bear is Grandfather, the Old Man, my kin, included within the circle of compassion such that the conflict of hunting them created the first art forms known, ceremoniously placed skulls and bones which served as ritual requests for forgiveness, found calcified in the limestone caves of Carpathia. Imagine setting out to kill and eat your Grandfather who also happens to be the messenger of the gods. Oh, and did I mention claws? Think Sumo wrestler with steak knives. You love him and revere him and want to eat him, if he doesn't eat you first.

The Princess is saved from being devoured by Bear because she begins to care about the fact that she doesn't care. Sad at this lack of love she wanders off to find them something to eat. She finds a dove with a broken wing and tends it, making a bandage from the hem of her expensive dress. Life is changed by having doubt thrown upon it. I once told my analyst I felt disillusioned. He said, 'oh, good.'

The Princess' sacrifice of her expensive dress is letting go of an identity centered in narcissistic entitlement. 'Out of the natural state of identity with what is mine there grows the ethical task of sacrificing that part of oneself which is identical with the gift . . . and the corresponding claim attached' (Jung, CW 11, para. 390). This happens as a direct result of her willingness to despair and become conscious of her ugliness. She then helps free a Giant Fish which the Dwarf has frozen in ice and later a Deer caught in a snow drift. All of these are functions of the psyche, either wounded or frozen or stuck. They can be liberated once the personality has faced and sacrificed its own inflation and vanity.

While the Princess is gone, the Dwarf wrecks her home and blames it on Bear; manipulating her to go back to her father with false stories of him being on the verge of death. When she reaches the castle, she realizes she's been tricked. More importantly, out in the garden, the Tree has responded to her new selflessness. It is ringing and singing at last – the beautiful Tree! Being willing to be depressed and anxious about the right things awakens love in her. The Tree knows this. The Princess' new kindness and self-sacrifice creates, 'that space of experiencing between the inner and outer worlds, and contributed to by both, in which primary creativity exists and can develop' (Winnicott, 1951, para. 17). Sacrifice which leads to involvement and participation in life rather than mere bartering for protection clears a sacred space, invites new possibility and engages a helpful response from the unconscious.

Now the Princess must get back to Bear whom she realizes is the Prince, but the Dwarf throws up a great barrier of thorns. She leaps over with the help of Deer, whom she rescued from the snow drift. Then he sends a flood, but Giant Fish comes to her aid. He drops her in a deep ravine but the Birds, whose friend had a broken wing, arrive to fly her out. Eventually, the Dwarf encircles the tree in flames, but the brave Princess calmly walks through them to embrace the Tree. Dwarf disappears and Bear is restored to his human form.

The house of Bear and Princess

Bear and Princess build a seemingly impossible house together. It's their turning point, their metanoia, because they co-operate despite brutish Bear behavior and the Princess' foul appearance. The house is a symbol of 'both/and' rather than 'either/or,' inner space you can stretch out in with wild and ugly, where meaning can be found 'in-between.' This is achieved by a significant shift in the Princess' attitude. She realizes she needs Bear. She begins to behave in a propitious way. She begins to listen, to pay attention. It's the kind of step you take when you begin to record your dreams or find yourself wanting to take environmental action. When Bear warns her that the project will take a thousand apron loads of rock to complete, she accepts without complaint. 'The essential feature of transitional phenomenon,' says psychoanalyst Donald Winnicott, 'is a quality in our attitude when we observe them' (Winnicott, in Praglin, 1974, pp. 81–9).

The significance of this attitude, of being committed to a co-operative venture irrespective of the time it takes, or the suffering involved is exemplified in the Zen tradition in a story called 'The Taste of Banzo's Sword':

> A man went to see Banzo and said, 'If I work very hard, how soon
>   can I be enlightened?'
> The Zen master looked him up and down and said, 'Ten years.'
> The fellow said, 'No, listen, I mean if I really work at it, how long?'
> The Zen master cut him off. 'I'm sorry. I misjudged. Twenty years.'
> 'Wait!' said the young man, 'You don't understand! I'm. . .'
> 'Thirty years,' said the Zen master.
>
> (Castro, 2013)

The Princess begins to care for whatever crosses her path and does so without thought of return or how long it will take. The seduction of hope, that every-thing will be OK tomorrow, gives way to longing which gives you a wheel you can put your shoulder to, today. It is these relations forged with her inner world – represented by Dove, Fish and Stag – which manage to keep her buoy-ant despite regressive forces preferring her to be dependent on outward powers. She is developing a new relationship with suffering, and has accepted that it is part of love.

She no longer shies away from it and so it doesn't hurt forever, as it used to.

While the Princess is rediscovering her connection with the wounded, the frozen and the stuck aspects of the psyche, the Evil Dwarf attacks and destroys the home she has made with Bear, much as our home/planet is being destroyed by the dark face of phallocentric extremism for whom power has become more important than progeny. Despite lordship over the castle, the King's identifica-tion with the topmost levels of the Psyche means Nature is now run through with everything hived off in order to reach such self-congratulatory heights, including the ambivalent ancient gods we outwardly deny but with which we nevertheless covertly battle. They 'have become diseases; Zeus no longer rules Olympus but rather the solar plexus, and produces curious specimens for the doctor's consult-ing room, or disorders of the brains of politicians and journalists who unwillingly let loose psychic epidemics on the world' (Jung, CW 13, para. 54).

When we talk about 'projection,' it is normally understood this is something which simply happens between people. Less obvious is our projection onto matter and ultimately onto Nature itself, which our civilized culture tends to view either as red in tooth and claw, as something violent, or as something mechanical with which wizardly engineering can then endlessly tinker. One of the earliest and most vivid examples of this paranoid projection onto Nature was the destruction of sacred groves by monotheistic kings in the Eastern Mediterranean during the seventh century BCE during the eradication of the Goddess religions, for whom trees were a symbol of and refuge for the Goddess herself. 'And [King] Josiah brake in pieces the images, and cut down the groves, and filled their places with the bones of men' (2 Kings 23:14). In the epic of Gilgamesh, the prototypical king cuts down the sacred grove of the Great Mother and makes his city gates out of them. Such attacks upon Nature are motivated by more than greed or mindsets of scarcity. They are more than unfortunate outcome, or collateral damage. They are symbolic events.

The King's legacy

When the King exploits his daughter for the Tree, he incurs something terrible upon both himself and the land. His ivory towered consciousness costs him connection to the Forest. It is now lorded over by the Evil Dwarf who can roam the paths and bridleways of the unconscious as he pleases. The King's new situation is far from the stable ground he hoped for. He is surrounded by threatening forces made more potent for his denial of their reality. He counters obliquely with efforts to master Wave, to conquer Mountain, to penetrate Jungle, to tame Plains and gouge the Land but, despite all these compensatory efforts, he still can't sleep at night for wanting to possess and own the Tree.

This belligerent attitude toward Nature has serious consequences. Forest is now animated with all the ancient gods the King prefers to ignore. They threaten to swallow up the castle with the King inside it, an unfortunate legacy exemplified by Doris Lessing's character Mary Turner in her dark novel *The Grass is Singing* (1950). A daughter of Empire in colonial Rhodesia, this Princess is carried off to madness by unconscious forces whose existence she prefers to deny, symbolized by her black servant and the bush from which he emerges. Everything from which she hoped to protect and distance herself, everything ugly, complicated or beyond her, simply materializes in the pulsing African forest.

Mary's un-lived life, everything which doesn't fit with being a Princess, is experienced as coming at her from outside, the malevolent sun, the hateful trees. She fantasizes how the forest will finally be her end, 'a branch would nudge through broken windowpanes, and slowly slowly the shoulders of the trees would press against the brick until at last it leaned and crumbled and fell' (ibid. p. 207). Her paranoia increases. She tries to keep an eye on the trees, so they do not sneak up on her or rush her unawares but even these precautions are not enough. Mary is attacked and killed. Her psyche finally collapses, and even though the agent of her demise seems to be a knife wielding third party, Lessing leaves us in no doubt about the true perpetrator. As the knife falls, she writes, 'and then the bush avenged itself. The trees advanced in a rush, like beasts, and the thunder was the noise of their coming' (ibid., p. 127). Mary is murdered from within.

E.M. Forster also describes this experience of Nature as a malevolent lover in *A Passage to India* (1924). The Princess of this story is the youthful Adela Quested, who has an experience in the mysterious Malabar caves which leads her to make a false rape allegation against her guide. Something violently carries her away which feels synonymous with a gross invasion of privacy but was in fact committed by something other than a mere man, some ineffable mystery whose sudden intrusion into consciousness felt like the unwanted advances of a demon lover.

Princess Quested's problem is in her surname. She thinks life is all over at twenty one. She's been there and done that. She knows everything. So, she trips into the caves with nothing more on her mind than her own boredom and is rudely surprised to discover the cave is occupied. What she is missing is the kind of propitious attitude exemplified by Mircea Eliade's description of indigenous Bakitara miners whose own ventures into the Earth are accompanied by a very different mind-set. They perform

> rites calling for a state of cleanliness, fasting, meditation, prayers and acts of worship. All these things were ordained by the very nature of the operation because the area to be entered is sacred and inviolable; the spirits reigning there are about to be disturbed; contact is to be made with something sacred.
>
> (Eliade, 1956, p. 56)

Becoming more consciousness really does take you away from what you think you know. The experience of the Self 'is always a defeat for the ego' (Jung, CW 14, para. 778). You are no longer in Kansas. So, of course the Princess resists the grown-up knowledge she thinks she wants. She is being asked to do more than change her mind. Something is unfolding that requires more than action. Rather, she is being asked to bear the experience of being acted upon.

Protection vs. dominion

The King's moral dilemma, how to keep the Tree for himself while making it all look like an accident, is complicated by a special feature of what it means to wear the crown. Kings may look like Paramount Chiefs, but there are important differences which have an impact on the story difficult to imagine. 'This was not simply a quantitative extension of a ranking system; it was a truly qualitative change by which society had entered a new realm' (Kirch, 2010, p. 17).

This new system entailed God-men. Kings are identified with and ordained by the gods. You get to be above the law. There are new rules about who you can kill without calling it murder, and new attitudes to the land. Not only does everyone become second-class citizens with fewer rights than before, so too are they divided against one another and against Nature with the infinitesimal isolating hierarchies which centralized power needs to maintain itself. The Chinese

Emperor Ming (28–75 AD) had twenty-three wives, the salient detail being that they all had different ranks, which meant they spent most of their time engaged in elaborate ritual displays of dominance and subservience while secretly trying to poison each other.

The Emperor's power depended on the inequality between himself and others but, more importantly, between the people themselves. The exploitation this gives rise to extends by association to the land which is now 'ground rather than earth' (Baring and Cashford, 1991, p. 495). Later than Ming, the Chinese sage Lao Tzu observed, 'If we stop looking for persons to put in power, there will be no more jealousies among the people' (1993, p. 1.3). We think of ourselves as consumers, forgetting the consumption responsible for devouring the planet is predicated on something darker and more fundamental than human greed.

In the Old Testament, Yahweh, jealous champion of kings, instructs Adam and Eve to subdue the natural world . . . 'and have dominion over the fish of the sea, and over the fowl of the air, and over every living thing that moveth upon the earth' (Genesis 1:26–8). This legacy has given rise to a civilization founded on cultural metaphors of Nature as an enemy to be conquered or a commodity to be traded, markedly different from the attitudes of indigenous consciousness which views the natural world as extended family.

Tradition has it the king generally comes to a bad end. Scandinavian kings ruled for twelve years, after which they were put to death or a substitute found to die in their place. Just the right kind of sacrifice might appease the gods, sacrifices made in their ones and twos all decked out in costumed finery to begin with. But then, maybe it would cover all the angles if they were also sacrificed in their uniformed millions . . . along with homes, gardens and countryside. When the King accidentally-on-purpose trades the Princess in exchange for the Singing, Ringing Tree, he's buying stability and time for himself in a devious and covert Faustian pact. Since the Princess' legacy is the realm itself, the King is now placed in an ambiguous relationship with his stewardship of the land. By association, it too can be sacrificed as a bargaining chip to possess the sacred Tree. Nature is devalued for as long as the Self is regarded as an object of consciousness by the King, who delegates sacrificial substitutes for the privilege of propitiating nameless gods.

The King thinks he is the hub of all things, so he's blinded to the Princess' predicament.

> Inflation magnifies the blind spot in the eye, and the more we are assimilated by the (Self), the greater becomes the tendency to identify with it. A clear symptom of this is our growing disinclination to take note of the reactions of the environment.
>
> (Jung, CW 9ii, para. 44)

This is quite apparent in our collective attitude toward climate change, perhaps best expressed by the myopic and Orwellian rollback of regulations by the US

Environmental Protection Agency which now facilitates pollution rather than preventing it. The sacrifice of resources and young lives in return for the King's prolonged grip on power is likewise being enacted by current US saber rattling with Iran, 120,000 troops now earmarked for deployment.

Revolutions of the soul

In the Gnostic tradition, you find descriptions of three very different types of consciousness. A parable explains. The simple person goes home wondering what is for tea. The complicated person goes home wondering about the contradictions of life and the vastness of the cosmos. The wise person goes home, like the simple person, wondering what is for tea. These three people represent distinctly different ways of being. The simple person thinks that the world is what they know of it, like the Princess at the beginning of our story for whom self-fulfillment was about achieving her own ends and filling her own belly. The complicated person has lost their appetite because s/he is plagued by all kinds of things which never bothered them before – Bears and Dwarves and ugly reflections. There is a sneaky feeling of not being master of your own house. In fact, you've found sudden yawning depths in yourself you had no inkling of when you got up, enough to make you feel perplexed and turned about.

The wise person has let go of trying to figure it all out but only by virtue of cohabiting with what is ancient in us and allowing the mystery of it all to be what it is, 'not as a cloak for ignorance but as an admission of his inability to translate what s/he knows into the speech of intellect' (Jung, CW 16, para. 110). What happens when your fear of looking within because of what you will find there becomes the fear of what you will become as a result? Not all change and growth are by steady increment. There are also paradigm shifts which change the very way you see things, let alone what it is that's on view. 'The merely natural man must die in part during his own lifetime. He will infallibly run into his Unconscious, a fatality he has no inkling of until it overtakes him' (Jung, CW 1, para. 471). This brush with the Self, if it were confined to an event replete with marquee and canapes, might be tolerable. We could give it a name and have an elaborate ceremony to try and contain its impact.

But it has a way of happening by itself in all kinds of unforeseen circumstances for which there are no rehearsals. Ritual, humanity's response to the gods, may go some way to contain the de-integrating effects of the Evil Dwarf, but it is in the nature of the Self to bust in on consciousness under its own steam, leaving the ego deflated and deposed in a way that can feel like abduction.

I was at a party once. Someone had prepped a very heavily made up young woman that I was a psychotherapist. She made a beeline for me and, without qualification, launched in:

> "I met a baby dragon in the woods. I took it home and looked after it. When it was grown, I released it into the woods again, but the villagers came up

and beat it to death!" By now she was crying loudly, the mascara streaming down her cheeks.

Everybody staring.

"What does a dragon mean?"

"Do you see," I asked, "how you just beat it to death?"

She didn't want to know the dragon. She wanted to know *about* the dragon. She wanted to turn it over in her hand like a trinket and so its aliveness was lost to her. Poor woman. Understanding the meaning of her experience was bound to constitute a revolution of the soul. The church fathers locked up Galileo for years after they accepted the math involved in his assertion that the Earth revolved around the sun and not the other way around for the same reason. They just couldn't accept the Earth was not the center of the universe and God might be involved elsewhere. Having tea or a quick nap. Their worlds were turned upside down. 'The personality becomes so vastly enlarged that the normal ego personality is almost extinguished' (Jung, CW 16, para. 472).

A man dreams he is out backpacking with friends. One by one, they disappear or are lost. Bit by bit, he loses his own way. Then his pack is gone. Then his boots. One item of clothing after another, until he is stumbling along naked in the dark. Eventually, through the trees, a growing light, a swelling raucous song. A clearing in the forest opens. Ecstatic dancers cavort about a vine-covered statue. The ground shakes. The statue is but the little finger of a Mighty Being buried in the bowels of the Earth. The vine is like a ring on its finger from which grapes the size of plums rains down while a ringing, singing voice invites him to eat.

'The Singing, Ringing Tree' requires us to shed what we think we know of ourselves rather than gathering up virtues or becoming anything. You must lose your golden cup and silver plate if not your way through the woods, along with all the other stuff you are identified with, in order to stumble across it. In Sumerian mythology, the goddess Innana descends to her dark and queenly sister Erishkigal; she is deprived of a layer of clothing at every gate and must arrive naked and bowed where she is killed and hung up on a meat hook. You wouldn't wish it upon your worst enemy, let alone undergo something similar yourself.

Sometimes the process of caring for concerns beyond the personality can feel like what the alchemists describe as the *mortificatio*, the dying, a fun process attended by what is cheerfully described as 'the torments' (Jung, CW 13, para. 439). This process of stripping away is a return to the Princess' original nature, and though the thousand apron loads of sharp stones cut into her, the Princess can bear it because her suffering now has meaning – she is involved in something bigger than herself.

The flame-encircled Tree

The Princess walks through the circle of flames and remembers her identity with Nature. Alchemically, flames burn the dross and impurities from the soul. The

King would burn because he is still inflated with flammable trappings and titles. The same idea is expressed slightly differently by canonical gospel, 'It is easier for a camel to go through the eye of a needle than for the rich man to enter the kingdom of Heaven' (Matthew 19:24).

Fortunately, the Princess has accepted her ugliness and developed a growing care for the world. Her co-operation with Bear has forged a sense of identity rooted in her own psyche which the flames cannot touch. The same idea is expressed in Norse mythology when Brunhild is encircled in flames for defying her father, Odin. She can only be rescued by a hero who loves her, whose relatedness transcends the limiting and partisan considerations of ego consciousness.

Such tests are kindly provided by the Evil Dwarf. As king of Tricksters, the Evil Dwarf is also Mercurius, famous for his dual nature as both bane and psychopomp of the individuation process. At first glance, he is a perverse contradiction, creating all kinds of suffering. Strangely, the Princess grows on account of all this evil. She is compelled to face her own ugliness and her own lack of care. Through the Evil Dwarf's impositions, she must become increasingly conscious. Via his frustrations of her efforts, she is compelled to develop a relationship with Bear, ultimately fruiting as a song of connection which has transformed her love of power into the power of love.

So, it is a mistake to go unseating Evil Dwarves or Weak Kings in a big show of gore and torch-bearing because you're liable to just get more of the same: 'the king must die; long live the king!' The trials faced by the Princess are not so much circumstances to be overcome by her will but as a result of having developed through her interaction with Bear a sense of relatedness with Dove, Fish and Deer, which then carry her through the Dwarf's trials unscathed.

What seems to be required is the kind of humble and forgiving co-operation expressed in the exchange between Princess and Bear. Having endured the shock of the encounter, there is a kind of flowering between them. Something opens. Bear responds constructively to the sacrifices which she makes. Something new emerges in the space created. From the Egyptian *Book of the Dead*, voices beyond the veil call for sacrifices: 'Give me my mouth, I want to talk. My two hands cling like ancestors. My lips are red as ox blood. Give me raisin cakes and beer. Bless me with ancient dreams. Give me songs green as earth' (Ellis, 1991, p. 47).

The gods hunger for sacrifices without which they will settle for your children and your countryside instead. The raisin cakes, or what have you, made with propitious intent, offered in quiet dignity, will do more than open inner space. It will inoculate against the compulsion to make such sacrifices unconsciously. Like Don Quixote, Western culture thinks of itself as evolved while enacting windmill tilting policies against the natural world it has imagined to be evil giants. The price of maintaining inflation permitted to parade as sophistication is a world of paranoia and denial then created. The intensification of 'Us and Them' in place of 'I and Thou' proliferates like plague once someone can be conned into being king for a day. The gods must be appeased for the privilege and for want of raisin cakes and ox blood, paint their lips with sap from the nation's youth and gouge great holes in the land.

The Tree's dark roots and aerial branches transcend opposites. It is the Self, expressed as a symbol of Nature. Nature, like the Self, is something we are made of and yet faced with. In an interior way, you have the deflating renunciation of thinking you are master of your own house. In the 'outer world,' it takes the shape of buoyant care for and identity with Nature which can feel like trial by fire but is an invitation across a threshold of consciousness into greater wholeness.

The Tree is also the Goddess herself. When the Princess passes through the flames, she does so as though reunited with the Great Mother. It's a religious experience. Deconstructing or dismissing the father/God religions is not enough. In order to be redeemed, she must remember and actively embrace the Divine Feminine whose principle of relatedness repairs our connection to Nature and to one another.

References

Baring, A. and Cashford, J. (1991). *The myth of the goddess*. London: Arkana.

Barks, C. (2002). *The soul of Rumi*. London: Harper One.

Castro, M. (2013). 'The taste of Banzo's Sword' in word press dad almighty.

Eliade, M. (1956). *The forge and the crucible*. Chicago, IL: Chicago Press.

Ellis, N. (1991). *Giving a mouth to Osiris*. Grand Rapids, MI: Phanes Press.

Forster, E.M. (1924). *A passage to India*. London: Edward Arnold.

Jung, C.G. (1953–77). *Except where indicated, references are by volume and paragraph number to the collected works of C. G. Jung*. 20 vol., ed. by H. Read, M. Fordham, and G. Adler, trans. by R.F.C. Hull. London: Routledge and Princeton: Princeton University Press.

Kirch, P. (2010). *How chiefs became kings*. Los Angeles, CA: UCLA Press. Available at: squin chpix.blogspot.com/2014/12/the-hawaiian-islands-case-study-in.html? view=flipcard.

Lao Tzu. (1993). *Tao te Ching*. London: Hackett Classics.

Lessing, D. (1950). *The grass is singing*. London: M. Joseph.

Mesocosm. (2012). *The bear, the raven and the shamanism of the pacific northwest*. Available at: mesocosm.net/2012/08/03/raven-and-shamanism/.

Praglin, L. (1974). The nature of in-between. *Journal for the Study of Interpersonal Processes*. Available at: universitas.uni.edu/archive/fall06/pdf/art_praglin.pdf.

Von Franz, M.L. (2001). *The golden ass of Lucius Apuleius*. Boston and London: Shambhala.

Winnicott, D. (1951). *Transitional objects and transitional phenomena*. Oxford: Oxford University Press.

Woodman, M. (2009). Available at: www.academia.edu/1188306/Sieff_D.F._2009_Con fronting_Death_Mother_An_interview_with_Marion_Woodman.

On Hearing News of a Pregnancy

So precarious is the thread of human life,
dear daughter, that I marvel at its extension
onward into the gigantic, unforetellable
room of the future. May the deep
goodness of Nature by its patient insistence
prevail against the drones –
against the bombs our lawless
freedoms have unleashed – and may the God
who escorted matter to these unlikely pinnacles
of human kindness and of human cruelty,
for a child's sake –
for this child's sake –
nudge with his weightless hand the wildly swinging
and fateful balances –

D.M. Black

Index

For Product Safety Concerns and Information please contact our EU
representative GPSR@taylorandfrancis.com
Taylor & Francis Verlag GmbH, Kaufingerstraße 24, 80331 München, Germany